I. L. ROZENTAL
Big Bang Big Bounce

I. L. Rozental

Big Bang
Big Bounce

How Particles and Fields Drive Cosmic Evolution

With 18 Figures

Springer-Verlag Berlin Heidelberg New York
London Paris Tokyo

Professor Dr. Iosif L. Rozental

Space Research Institute, Academy of Sciences, USSR, Profsojusnaja, 84/32
SU-117810 Moscow, USSR

Translator:

Dr. Juri Estrin

Vogelbeerenweg 6, D-2110 Buchholz i. d. N., Fed. Rep. of Germany

Title of the original Russian editions:
Elementarnye chastitsy i struktura bselennoj © "Nauka" Publishing House, Moscow 1984
Problemy nachala i kontsa metagalaktiki © "Znanie" Publishing House, Moscow 1985

ISBN-13:978-3-540-17904-7   e-ISBN-13:978-3-642-72745-0
DOI: 10.1007/978-3-642-72745-0

Library of Congress Cataloging-in-Publication Data. Rozental, I.L. (Iosif Leonidovich) Big Bang Big Bounce. Based on: Elementarnye chastitsy i struktura Vselennoï, and Problemy nachala i kontsa metagalaktiki. 1. Cosmology. 2. Particles (Nuclear physics). 3. Astrophysics. I. Title. QB981.R73 1987 523.1 87-16457

2153/3150-543210

Foreword

In a foreword, an author usually elucidates the aim of his book and describes an idealized reader to whom it is addressed. The first task — the formulation of the scope of the book — is the easier one, for the second one involves assessing a reader's personality, and no "specification" should warrant the author's being accused of snobbery, underestimating the reader, or other sins of that kind.

It is natural to commence with the first task. The last two decades have been marked by extreme, albeit somewhat unexpected, progress in the unifying approaches to fundamental physical theories. During the same time, a reasonably consistent picture of the early stages in the evolution of the Universe, starting from the time $\sim 1\,\mathrm{s}$ reckoned from the beginning of its inflation, began to take shape. These questions have been separately treated at very different levels; their systematic presentation is the subject of monographs, sometimes very solid ones, containing many formulas not tractable for a layman.

A more recent subject, which has been not given enough consideration in the popular scientific literature, concerns the results obtained toward a synthesis of the microcosm and the macrocosm. This approach reveals common features of elementary particle physics and cosmology. An unexpected result of this symbiosis is the hope of understanding the nature of the fundamental physical constants — a problem which appeared to be unsolvable until recently. The structure of the Universe is unstable with respect to the magnitude of the fundamental constants in the sense that a small variation in their values would lead to a major qualitative change in the structure of the Universe.

In accordance with the general trends in fundamental physics, this book consists of four parts. In the first two chapters, the unification tendencies in elementary particle physics and the advancement of cosmology are considered. The third chapter deals with a synthesis of elementary particle physics and cosmology. Of course, not all elements of the synthesis are discussed: Some are too hypothetical; others lie beyond the sphere of the author's interests. The fourth chapter is concerned with the formation and decay of the superuniverse.

Now it is finally time to tackle the second, and more ticklish, task, namely sketching an "idealized" reader of this book. Not being a professional popularizer, the author undertook to keep to the beaten track in order to solve this problem: he consulted the recognized examples. One such example is a recent book by the renowned physicist and popularizer, a Nobel Prizewinner, S. Weinberg, entitled *The First Three Minutes*. Weinberg writes in the foreword: "I have written for one who is willing to puzzle through some detailed arguments, but who is not at home in either mathematics or physics. Although I must introduce some fairly complicated scientific ideas, no mathematics is used in the body of the book beyond arithmetic, and little or no knowledge of physics or astronomy is assumed in advance".

The First Three Minutes is composed in the following way: there are no mathematical formulas in the main text; all of them are contained in a supplement. An Appendix provides a glossary of special terminology. It explains, in particular, the notions of a neutron, a proton, and an electron. Acquaintance with this captivating book made it clear to me that my subject cannot be presented following the same pattern. The structure of Weinberg's book is perhaps acceptable for the presentation of his particular topic — the evolution of the early Universe. It does not, however, seem suited for dealing with the broader subject of synthesizing modern trends in physics and cosmology, embracing the advances in physics achieved in the last decades, and outlining certain future prospects.

Furthermore, it can be explained to an attentive reader unfamiliar with the notion of a proton what proton decay is, but it appears virtually impossible to demonstrate its impressive significance for the progress of fundamental physics. Emphatic attributes would have to be used, like "the experiment of the century," "the experiment of the millenium," etc. The author is not inclined to take this path, for it implies underestimating the reader and a limitedness of goal, namely, eliminating the background against which a particular, concrete fact manifests itself.

Introducing a glossary does not seem to be justified, either. The author is sure that if there were an odd man not familiar with Shakespeare and the English language who wished to read the original of *Hamlet* with the aid of a dictionary, he would not agree with the high appraisal of this work of literature. For such a person, *Hamlet* would occur to be a commonplace play about a palace coup d'état. The whole effort of the reader would be devoted to understanding the plot, and he would hardly be aware of Hamlet's titanic struggle with himself and with the ill fate that culminates so tragically. Neither would he appreciate the perfect literary form in which the plot is cast. The monologue "To be, or not to be" should be read by a person who has a perfect command of the English language. However, a reader not having this virtue has an alternative: to read "To be, or not to be" in an excellent translation.

Unfortunately, to understand the fascinating but very profound modern physical ideas, no such solution is available. It is not possible to eliminate, without considerable damage, the mathematics by just "translating" it into a common, colloquial language. Mathematics that used to be an obedient tool in the hands of the physicists has long become a definite way of thinking.

An anecdote comes to mind in this connection. J. W. Gibbs — one of the founders of statistical physics — was a member of the Scientific Board of Yale University. Usually Gibbs did not participate in debates. But once a heated controversy arose about what is of more importance: teaching languages or mathematics. Gibbs was urged to give his opinion on this matter; his answer was: "Mathematics *is* a language." Time has confirmed the extreme actuality of this aphoristic statement.

With these considerations in mind, the author believes that the only way to avoid profanation of the subject of the book is to share the burden of responsibility with the reader. It is assumed that the reader has taken courses in physics and mathematics at the undergraduate or graduate level, and that there is no need to explain here what a proton or an electron is all about. It should be stressed that a limited use of mathematics does not allow rigorous proof of all the statements made in the book; the reader will simply have to believe some of them.

It goes without saying that the author cannot claim his presentation of the questions considered to be impeccable. He would be grateful to the readers willing to communicate their comments and suggestions to him.

I thank A.D. Linde, M.I. Podgorezkii and V.V. Usov for fruitful discussions on the questions touched in this book.

Preface to the English Edition

Some ten years ago, cosmology seemed to be a settled discipline: the Friedmann model explained all observational facts pertaining to the Universe regarded as an unique entity; the elegant picture of a Big Bang in which the Universe emerges from "nothing" satisfied the aesthetic taste of the majority of the experts. Quite common was the opinion that the Universe is the only object and that it is senseless to even speak of a space-time beyond the Universe.

However, in the late 1970s new concepts obviously contradicting these ideas started taking shape. In the first place, the anthropic principle should be mentioned. It questioned the literal interpretation of Copernicus' conception of the Earth as an ordinary planet. Indeed, from the viewpoint of physics, the Earth occupies an ordinary place among the planets. It is however, the only dwelling of civilization, i.e., of highly developed biological (chemical) forms of matter. Their occurrence requires very specific physical conditions, placing strong limitations on the arbitrariness of the physical laws.

Furthermore, an enormous instability of the structure of the Universe with respect to the numerical values of the fundamental constants was discovered. It turned out that in a sense, the observable Universe has a fluctuational character. The values of the fundamental constants that make the existence of complex forms of matter possible are strongly distinct from their analogues, well studied in the laboratory.

The only way to account for all these facts in the framework of physical concepts, without invoking a metaphysical power, is to admit the existence of an (infinitely large?) number of universes: metagalaxies. Furthermore, one has to assume that the physical laws governing our Universe result from a "random" choice, dictated by the necessity to account for the occurrence of complex forms of matter in the Metagalaxy.

This range of subjects was dealt with in the book, *Elementary Particles and the Structure of the Universe*, written (in Russian) by the author in 1982. The question of the physical modelling of the origin of many universes (metagalaxies) remained unclear, however. Between 1981 and 1983, an extensive series of works appeared on the formation of the superuniverse and its decay within a time of $\sim 10^{-35}$ s into many metagalaxies governed by their own laws (cf. the Inflationary Universe model proposed by A. Guth, A. Linde, and A. Starobinskii). This is the central subject of Chap. 4 "The Beginning and End of the Metagalaxy", written in 1984.

It should be emphasized that progress in cosmology was not limited to the two subjects mentioned. From 1983–1985, a flood of papers appeared, concerned with the analysis of the nature of physical space. The fundamental idea underlying this new direction is gauge invariance and its geometric interpretation on the basis of the Kaluza-Klein model. This model, to which Einstein devoted the last years of his scientific activities, gained momentum as a result of general recognition of the pre-eminence of the gauge invariance principle.

New profound physical and mathematical ideas proposed by many outstanding scientists (S. Weinberg, E. Witten, S. Hawking, B. de Witt, and many others) suggest that the dimension of initial physical space was larger than that observed ($N = 3$). This real dimension is $N \geq 10$. Spontaneous compactification results in a reduction of the dimension, leading eventually to $N = 3$ for our Metagalaxy.

This important and interesting problem is only touched upon in the treatise presented. Its detailed consideration would require a separate book; the author hopes to write such a book in the future.

I am very grateful to Mrs. V. Dittrich who edited the English edition of this book, for her guidance and competent assistance in the preparation. In conclusion, the author would like to thank Springer-Verlag for the opportunity of presenting his viewpoint concerning the origin of the Metagalaxy to readers in the West.

Moscow, August 1987 I. ROZENTAL

Contents

1. **Elementary Particles** 1
   1.1 Fundamental Interactions 1
   1.2 Quantum Numbers of Elementary Particles 5
       1.2.1 Spin 5
       1.2.2 Isospin 6
       1.2.3 Strangeness 8
       1.2.4 Color 9
   1.3 Basics of Classification of Elementary Particles 11
   1.4 How Elementary Particles Interact 12
   1.5 Unified Field Theories 14
       1.5.1 The Universal Constant 15
       1.5.2 Unified Symmetry 18
   1.6 Proton Decay 21

2. **The Universe** 27
   2.1 A Bit of History 27
   2.2 Friedmann's Model of the Universe 30
   2.3 Evolution of the Universe: A Quantitative Analysis 31
   2.4 The Universe: Open or Closed? 35
   2.5 A Hot Universe 36
   2.6 Baryonic Asymmetry of the Universe 39
   2.7 Cosmologic Nucleosynethesis of Helium 44
   2.8 The Origin of Galaxies 48
   2.9 Stars 53
       2.9.1 Classification of Stars 53
       2.9.2 Biography of a Star 56

3. **The Universe and the Elementary Particles** 61
   3.1 On the Relation between the Characteristic of Stars
       and of the Elementary Particles 61
       3.1.1 Stars and Planets: A Distinction 61

3.1.2   Stellar Parameters: A Quantitative Evaluation   62

3.2   Structure of the Universe and the Mass of the
Elementary Particles   66

     3.2.1   The Mass of the Electron   67

     3.2.2   The Mass of the Nucleon   70

3.3   Structure of the Universe and the Fundamental
Interaction Constants....................................   72

     3.3.1   The Strong Interaction   72

     3.3.2   The Electromagnetic Interaction   74

     3.3.3   The Weak Interaction   76

     3.3.4   The Gravitational Interaction......................   77

3.4   The Dimension of Space   79

3.5   Structure of the Universe and Quantum Numbers
of Elementary Particles..................................   82

3.6   The Anthropic Principle..................................   83

3.7   On the Numerical Values of the Fundamental Constants ...   85

3.8   Conclusion   88

     3.8.1   Defining the Words "Universe" and "Metagalaxy" ...   88

     3.8.2   Metagalaxy Formation.............................   90

4.   **The Beginning and End of the Metagalaxy**   97

4.1   Updating our Knowledge of the Metagalaxy...............   97

4.2   Describing the Metagalaxy   97

4.3   The Universality of the Physical Laws   98

4.4   The Very Beginning   99

4.5   Models of the Metagalaxy   100

4.6   The Friedmann Model.....................................   103

     4.6.1   The Long Way to Recognition   103

     4.6.2   Difficulties.......................................   105

4.7   The Physical Vacuum   107

4.8   The de Sitter Model: The Beginning of the Metagalaxy   110

4.9   The Structure of the Metagalaxy
and the Fundamental Constants   113

4.10   The Metagalaxy as a Fluctuation.........................   115

4.11   The Anthropic Principle   117

     4.11.1   Definition..   117

     4.11.2   Applications   118

4.12   The Birth of the Metagalaxy and of Metagalaxies..........   120

     4.12.1   Progress in Cosmology Brings Understanding......   122

4.13   Future of the Metagalaxy   124

     4.13.1   Open Metagalaxy   125

     4.13.2   Closed Metagalaxy   128

The most improbable things use to
turn out to be the most logical ones.
E.M. Remarque

1.Elementary Particles

1.1 Fundamental Interactions

There are four types of fundamental interactions whose existence is well
established. Most studied are two of them: the gravitational and the electromagnetic interactions. The foundations of the classical (nonquantum)
theory of the two interaction types were laid long ago (Newton, Einstein,
and Maxwell); they are generally known from physics textbooks.

Gravitational interaction that governs the motion of celestial bodies
and earth's attraction is characterized by Newton's constant $G = 6.7 \times 10^{-8}$
$g^{-1} cm^3 s^{-2}$. An excellent approximation that describes the gravitational
interaction of two point masses m, a distance r apart, is Newton's formula,

$$F = \frac{Gm^2}{r^2} \quad .$$

Electromagnetic interaction determines the motion of charged bodies. In
the general case, their law of motion is described by the Maxwell-Lorentz
equations. In the quasistatic approximation, an analogue of Newton's law,
the Coulomb approximation

$$F = \frac{e^2}{r^2} \quad ,$$

proves to work very well, though. (Here, e denotes the charge of each point
mass.) The magnitudes of Gm^2 and e^2 depend on the choice of the system
of units; this is of a hindrance when analyzing the common ground of the
two interactions. To facilitate comparison and problem solving in the framework of quantum field theory, one combines these quantities with universal
constants, viz. the Planck constant $\hbar$ and the velocity of light c, to get
dimensionless constants. Thus, the nondimensional gravitational constant
$\alpha_g = Gm^2/\hbar c$ and the nondimensional electromagnetic coupling constant
$\alpha_e = e^2/\hbar c$ are obtained, $e \cong 10^{-19} C$ being the electron (proton) charge.

1

It should be noted that there is a difference in the definition of the two constants, α_e being in a way more universal than α_g. Indeed, the definition of the number α_e contains the fundamental constants only, whereas the constant α_g involves a mass m which is, generally speaking, arbitrary. To eliminate this arbitrariness, it is common to fix the value of m by setting it equal to the proton mass m_p. This choice is quite natural, for the proton is one of the two stable particles constituting the bodies of the Universe; the other one is the electron, with mass m_e. The choice between m_p and m_e is a matter of convention. In the rare cases in which it is of some physical significance, the difference will be emphasized.

Let us now describe the properties of the two other interactions discovered as late as in the 20th century. One of them, namely, the *weak interaction*, governs the decay of particles into lighter ones. Historically, the first decay discovered was the decay of a neutron within an atomic nucleus (the β-decay), according to the reaction

$$n \to p + e^- + \overline{\nu} \tag{1.1}$$

(n, p, e^-, and $\overline{\nu}$ standing for a neutron, a proton, an electron, and an antineutrino, respectively).

Later on, it became clear that neutron decay is not unique: new elementary particles were successively discovered. This process has been intensified by progress in the development of accelerators. It turned out that all newly discovered particles have a common property: Heavy particles decay into lighter ones. Numerous investigations led to the conclusion that many decays are controlled by a unique interaction, referred to as the weak interaction, which is characterized by the Fermi coupling constant $g_F \cong 10^{-49}\,\mathrm{erg\,cm^3}$. The corresponding dimensionless coupling constant for the weak interaction is $\alpha_w = g_F m^2 c/\hbar^3$. The processes of collisions of neutrinos with matter are determined by the weak interaction as well.

The situation involving the *strong interaction* is not so clear. Some 10–15 years ago, the strong interaction was identified with the nuclear interaction that determines the state of protons and neutrons in atomic nuclei. However, the attempts to develop a passably consistent theory of nuclear interaction were not successful. Currently, one must be satisfied with a phenomenological description of the interaction between protons and neutrons, based on potentials whose form is fitted so as to match the experimental data in an optimal way.

A breakthrough seemed to be achieved with the progress of the dynamical theory of quark systems that led to the advent of quantum chromodynamics. In that scheme, the nuclear interaction was identified with the interaction in many-quark systems. However, the hope that a self-consistent theory of nuclear interaction can be constructed on the basis of the quark model, raised in the early seventies, has not been realized so far.

It is useful and instructive to trace the evolution of the quark interpretation of nuclear interaction. To do so, we digress, briefly outlining the quark model proposed by Gell-Mann and Zweig in 1964. According to this model, each proton and neutron consists of three point-like particles which are referred to as quarks and possess a charge that is a fraction of the electron charge e, viz. $\pm\frac{1}{3}e$ or $\pm\frac{2}{3}e$. This theoretical conclusion was seemingly in contradiction to the experimental evidence that all the observable elementary particles have an integer electric charge. Nevertheless, numerous experimental confirmations of the quark hypothesis (such as systematics of the elementary particles, the magnitude of the magnetic moments, the ratios of the interaction cross-sections, etc.) suggested that it deserves serious consideration.

But then a profound question arose: How can the existence of quarks be reconciled with their nonobservability in direct experiments? It should definitely be realized that at present, this problem, referred to as that of *quark confinement,* is far from being solved strictly. To date, a postulate is invoked which has rather a character of an incantation: "Quarks do exist, but in a bound state." Even though no solution of the confinement problem is available, one bases some expectations on the construction of a mathematical model that claims to provide a theory of the interaction between the quarks. It is this interaction that is identified with the strong interaction which, in the last analysis, causes nuclear interaction (see below).

Another digression is in order at this point. In 1954, Yang and Mills proposed a theory which is basically different from electrodynamics, but accounts for the interaction caused by the transfer of zero-mass particles. The only particle known at that time was the photon. The photon is the particle underlying electrodynamics (see Sect. 1.4 for details). That is why the Yang-Mills theory was considered just an exotic mathematical exercise.

The picture changed radically, however, when a need emerged for a theory describing the dynamics of quarks. It seemed natural to consider the massless particles introduced by Yang and Mills to be responsible for the quark interaction. These particles were named *gluons*; by analogy with quantum electrodynamics, one of the variants of the Yang-Mills theory is referred to as *quantum chromodynamics*. In the early seventies, when this idea ripened, the Yang-Mills equations were subjected to more scrutiny. As a result, the constant α_s was found to exhibit quite remarkable behavior, as distinct from quantum electrodynamics. This constant determines the quark-quark interaction which is currently believed to be the true strong interaction. It should be remembered (cf. Fig. 1.1) that from the viewpoint of contemporary field theory, the interaction is mediated by particles, i.e., *quanta* of the corresponding field.

3

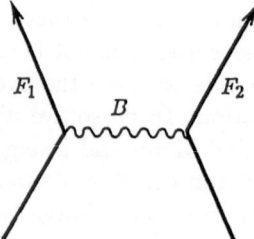

Fig. 1.1. Schematic representation of the interaction between the particles F_1 and F_2. The interaction is mediated by the particle B. It carries energy momentum and mass, as well as internal quantum numbers

Obviously, energy-momentum and hence − according to the special theory of relativity − mass is transferred along with a quantum. Elaborate calculations have demonstrated that the strong interaction coupling constant α_s essentially depends on the energy-momentum and the mass m transferred.

In a way, one had encountered a mass dependence of the constants α before (e.g., α_g and α_w), but quantum chromodynamics introduces a basic difference. In this theory, the dependence $\alpha_s(m)$ is deduced from quantum field theory, and not postulated, as was done earlier for the constants α_g and α_w on the basis of dimensional considerations. In addition, the variation of the constant α_s with the mass m has a specific feature: α_s decreases with increasing m.

It should be remarked here that the terminology repeatedly used above might appear contradictory. On the one hand, we speak of the constants α; on the other hand, we keep stressing their dependence on m. In fact, the constants α are only constant at a fixed m; they vary with changing m. That is why they are referred to as "running" constants. On this stipulation, the final expression for the dependence of α_s on m reads, in the asymptotic approximation when $m \gg m_p$:

$$\alpha_s \sim \frac{a}{\ln(m/m_p)} \ . \tag{1.2}$$

The quantity a depends on N_q, the number of the sorts of quarks. In a standard theory ($N_q = 6$), $a \sim 1$. It follows from this formula that $\alpha_s \to 0$ as $m \to \infty$. This is the famous phenomenon of *asymptotic freedom*. A similar dependence also follows from a more exact expression. Unfortunately, the latter has been also obtained by methods whose validity breaks down for $m \lesssim m_p$. A "true" expression for α_s at small m is missing, owing to the very fact that α_s is large, thus rendering standard computation techniques inapplicable. One can only state that for small characteristic mass m, corresponding to the proton (or neutron) size $r_N \sim 10^{-13}$cm, the coupling constant is large. This circumstance lends promise to the hope that the problem of quark confinement can be solved. Unfortunately, hopes often take quite some time to be before realized. This also applies to the case under consideration: the complex and important problem of quark confinement is not yet ultimately solved.

Furthermore, a rapid increase of the constant α_s with r approaching r_N inhibits progress in solving another problem, namely, that of nuclear forces. As a matter of fact, the advances in quantum chromodynamics did not bring about a better understanding of the nature of the nuclear forces. Calculations aiming at an interpretation of the experimental data are based on the use of phenomenological potentials. For example, in describing the interaction between two protons (or neutrons) or between a proton and a neutron, a potential is used that reads, in a somewhat simplified form,

$$V = \alpha_s m_\pi c^2 \exp\left(\frac{-r}{r_N}\right) \quad , \tag{1.3}$$

where m_π is the pion mass and $\alpha_s \sim 1$.

1.2 Quantum Numbers of Elementary Particles

1.2.1 Spin

An elementary particle is completely determined by the totality of its internal quantum numbers. The term "internal" emphasizes the fact that the quantum numbers refer to the particle itself; they are not related to uniformity of time or uniformity and isotropy of space, properties predetermining conservation of such familiar quantities as energy, momentum, and angular momentum. The best known and most familiar characteristics of elementary particles are their mass and electric charge. Coming from classical physics, they were adopted by elementary particle physics; their appearance was not connected with the development of quantum mechanics.

Principally new was the introduction of quantum numbers proper. Here, the particle spin should be named first.[1] Initially the spin entered physics

---

[1] In this section, we confine ourselves to the definition of those quantum numbers most essential with regard to the subject of this book.

as an intrinsic angular momentum M_e of an electron, its value being $M_e = \hbar/2$. However, such an interpretation of spin is severely wanting; in fact, according to the contemporary picture, the size of a structureless elementary particle (in which category the electron falls) is zero. On the other hand, the angular momentum is given by $M_e = [r_e p_e]$, p_e denoting the momentum of the particle. Since $r_e = 0$, the angular momentum is zero as well, $M_e = 0$, so that it cannot equal $\hbar/2$. Hence, the notion of spin as an attribute of an electron rotating in space is not justified, even though it might be quite illustrative.

The interpretation of spin in the quantum mechanical framework appears to be more consistent, taking into account the fact that the spin state of a system (an electron in the case under consideration) is defined by a vector in a certain space. Then the length (i.e., the norm) of the vector is specified in such a way that its projection onto one of the axes be $\pm\hbar/2$.[2] The spin state of the electron is determined by the probability that the spin projection will have a definite sign. One difference between a usual vector and the vector characterizing the spin should be emphasized, however: the latter is defined only by rotation about a fixed origin, whereas a usual vector can, in addition, be displaced in space.

A very popular definition of spin is based on group theory[3] where it is considered an irreducible representation of the rotation group SU(2).[4] Unfortunately, this definition can be characterized by a joke told by a hero of the Patriotic War of 1812, General A.P. Yermolov: "Every word in this sentence sounds foreign." To those familiar with group theory, the definition will appear trivial; to those unfamiliar with group theory, it is incomprehensible. In any case, it is most important to note that spin is a quantity which has the dimension of angular momentum; summation of these two spins obeys the standard quantum mechanical rules.

Now let us turn to other quantum numbers, not related to customary physical space.

1.2.2 Isospin

In 1932, Heisenberg noticed a striking proximity of the proton mass, $m_p = 938.3\,\text{MeV}$, and the neutron mass, $m_n = 939.6\,\text{MeV}$.[5] In this connection,

---

[2] It should be noted that even though the spin of an elementary particle may differ from $\hbar/2$, it is always a multiple of this quantity.

[3] Group theory is unfortunately beyond the scope of this book, and we have to refer to the following monographs: E.P. Wigner: *Group Theory* (Academic Press, N.Y., 1959) and J. Adams: *Lectures on Lie Groups*, (Benjamin, N.Y., 1969).

[4] The abbreviation SU refers to "special unitary."

[5] Henceforth, the masses of elementary particles are given in units of energy. It should be remembered that $1\,\text{MeV} \sim 10^{-6}$ erg corresponds approximately to the mass of $2 \cdot 10^{-27}$ g. The pion mass in these units is $m_\pi \sim 140\,\text{MeV}$.

in the same year he propounded an idea no less significant than his uncertainty principle — at least as far as elementary particles are concerned. He assumed that the proton and the neutron are two different states of one and the same particle which he named the nucleon. The observed difference between the proton state and the neutron state of the nucleon consists, according to Heisenberg's concept, in the difference of the electric charge ($e_p = e, e_n = 0$) that causes a small difference between m_p and m_n. The establishment of this fact alone would not have left a trace in physics. But Heisenberg did formulate a quantum mechanical interpretation of this phenomenon which — certainly in its significantly extended form — underlies the theory of elementary particles.

This concept is based on the following idea: the distinction between the proton and the neutron state is characterized by a new internal quantum number which Heisenberg called the isospin. The word "spin" in this notion emphasizes that the mathematical method of description of the isospin is related to that describing common spin.

There exists an isospin vector I whose projection, I_z, on the z-axis can take two values, $\pm\frac{1}{2}$, in accordance with the values of the projection of common spin. By convention, the value $I_z = \frac{1}{2}$ corresponds to the proton state and $I_z = -\frac{1}{2}$, to the neutron state of a nucleon. In the framework of this concept, the following question arises, however: In what space does the isospin vector exist? As discussed above, the common spin vector can be associated with common physical space, for spin and angular momentum have the same dimension. In the concept of isospin this association is completely lost. Isospin space is an abstract space in the sense that it is not connected with the physical space in which macroscopic bodies exist and in which the dynamic laws describing them apply. Isospin is an internal quantum number whose description requires introducing a special mathematical ("imaginary") space.[6] The basic significance of the idea of isospin was that for the first time, a nonphysical space (namely, that in which the isospin vector is defined) was introduced into elementary particle physics of necessity, and not just for computational convenience. The isospin method has found excellent confirmation in interpreting numerous experiments, and has served as a prototype for the description of the set of further quantum numbers.

In conclusion, two remarks should be made:

i)   The concept of isospin leads to two quantities to be conserved in strong interactions: the absolute value of the isospin vector and the

---

[6] Physical space is understood here as macroscopic space. In the framework of Newtonian mechanics, it is the three-dimensional Euclidian space, while in the special theory of relativity, it is the four-dimensional Minkowski space.

total projection I_z of a system of elementary particles (by analogy with common spin).

ii) The space in which isospin is defined is a two-dimensional complex Euclidean space. It can be thought of as a two-dimensional space in which each point x represents a complex number. The property of being Euclidean means that the modulus of a vector (x_1, x_2) in this space is determined by the sum $x_1^2 + x_2^2$.

1.2.3 Strangeness

In the early 1950s, the following phenomenon was discovered: The newly detected elementary particles (the K-meson and the Λ-particle) were never generated separately. For instance, the reaction $\pi^- + p \rightarrow \Lambda^0 + \pi^0$ (the π-pion) that did not seem to be forbidden by any physical law did not occur in experiments. On the other hand, such a reaction as, e.g., $\pi + p \rightarrow \Lambda^0 + K^0$ was observed perfectly well. There is a "golden rule" in elementary particle physics: Anything that is not forbidden should occur in Nature. There is no such categorical rule in macroscopic physics. A categorical negation in microphysics implies the existence of an exclusion rule. For this reason, the absence of reactions leading to creation of a single Λ-particle and compulsory generation of pairs thereof had to be reflected in a certain rule. This rule was formulated in 1952 by Gell-Mann and Nishijima. It attributed a new quantum number, called *strangeness* (S), to the K- and Λ-particles; this quantum number is strictly conserved in strong interactions. For simplicity, it was suggested that strangeness be characterized by integers. For the Λ-particle, $S = -1$ while for the K-mesons, $S = +1$; strangeness of the nucleons and the pions is zero. Thus, a "strange" behavior of new elementary particles is easily explained. In reactions involving generation of single Λ-particles, strangeness is not conserved: such a reaction is thus forbidden. For generation of Λ- or K-particles in pairs, strangeness is compensated for so that such reactions are allowed.

It should be noted that later on, elementary particles with larger absolute values of strangeness were detected. A general limitation on the magnitude of strangeness is $|s| \leq 3$.

At the next stage, a unified description of three conserved quantities, I, I_z, and S, became necessary. No consistent method of reconstructing the symmetry of dynamic equations from the conserved quantities is available. Still, a simple generalization of the isospin technique has proved to be very fruitful. It was suggested that all three conserved quantities be described by using, instead of the two-dimensional space of isospin, a three dimensional complex Euclidean space. The vector corresponding to all the quantum numbers is defined in this space.[7]

1.2.4 Color

Among the quantum numbers, the electric charge of the elementary particles plays a special role. On the one hand, it is a typically quantum mechanical number. All charged elementary particles have a charge which is a multiple of the quantity e. The introduction of quarks, with their fractional charge, does not essentially change this situation, one only has to consider the electric charge of the quarks as a charge unit (charge "quantum").

On the other hand, the electric charge also has another functional meaning. Being a characteristic of the electromagnetic interaction which determines the coupling constant α_e, the electric charge thus determines this interaction itself. It is this dual role that distinguishes electric charge from mass, for that matter: the latter quantity is not of quantum mechanical character, for there is no such thing as a "mass quantum."

The strong interaction is an essentially microscopic interaction. It would be natural, therfore, to conjecture, by analogy with the electromagnetic interaction, that a quantum number, the "strong charge", exists. However, the strong interactions of nucleons are not long range: "strong charges" do not manifest themselves at great distances. Hence, the "strong charge" hypothesis leads to the conclusion (again by analogy with the nonobservability of quarks) that the "strong charges" of the three quarks are mutually compensated for within a nucleon so that the net "strong charge" of a nucleon is zero.

One would think, by analogy with electromagnetism, that the simplest variant of such a hypothesis is a "strong charge" with three values, ± 1 and 0. However, this simplistic assumption would be in contradiction to the established symmetry considerations with regard to particles with charges of opposite sign. According to modern theory, systems in which the electrons (with charge $-e$) are replaced by positrons (with charge $+e$) are equivalent (the so-called charge invariance). Obviously, there is no such equivalence if charged particles are replaced by neutral ones. Therefore the characterization of the "strong charge" requires devising a quantity completely equivalent in all three modifications; in addition, the sum of the threee different charges must vanish, since the "strong charge," as distinct from the electromagnetic charge, has never been observed.

---

[7]   In terms of group theory, such a description corresponds to the transformation group SU (3), while (iso)spin symmetry corresponds to the group SU (2).

Both requirements mentioned ruled out the use of simple mathematical objects analogous to those in electromagnetism, for which a vast bulk of data were described by using the double-valued scale of real numbers. For that reason, a phyiscal entity, *color*, was chosen as an analogue and term for the "strong charge." It is well known that three colors (e.g., red, yellow, and blue) possess the property of complementarity, yielding, as a mixture, the color white. The color white — a symbol of achromaticity — corresponds to the absence of the "strong charge" in the observed elementary particles. The quarks in nucleons do possess a "strong charge", namely, "color." The combination of colors of the quarks contained in a particle is such that it becomes colorless, i.e., the "strong charge" of the particle vanishes. The field of a nucleon, i.e., the net field of the constituent quarks, rapidly decreases with distance. It is similar to the field of an electric multipole responsible for the van der Waals interaction. (The field from a dipole is inversely proportional to the cube of the distance.) It should be emphasized that the terminological association of the notion of "color" with the optical colors does not imply any relation to them.

The symmetry of quarks with respect to the three colors, as distinct from the symmetry of electric charges with respect to the plus and minus signs, leads to another important innovation. The particles mediating the interaction between the quarks, which are referred to as *gluons* (cf. the scheme represented in Fig. 1.1), possess color and thus transfer it, changing the color, but not the kind, of a quark. In such an interaction, a "red" quark, for instance, transforms into a "yellow" one. This reflects a fundamental difference between quantum chromodynamics[8] and quantum electrodynamics where the interaction is mediated by photons — electrically neutral particles which do not change the electric charge characteristic, i.e., its sign.

The occurrence of color in gluons results in an interaction between them that makes the equations of quantum chromodynamics fundamentally differ from those of quantum electrodynamics. The equations pertaining to electromagnetism are usually linear, whereas the equations of quantum chromodynamics are essentially nonlinear. The latter circumstance significantly complicates their solution. In particular, the problem of quark confinement mentioned above is connected with the nonlinearity of the equations of quantum chromodynamics that describe the quark interaction.

---

[8] In this way, the etymology of this notion becomes clear (chromatic being synonymous with colored).

1.3 Basics of Classification of Elementary Particles

Elementary particles are classified with respect to various parameters. A most general ground or particle classification appears to be provided by the value of the spin, s. Indeed, the behaviour of particles depends on whether their spin is characterized by an integer (0, 1, 2, ...) or a half-integer ($\frac{1}{2}$, $\frac{3}{2}$, $\frac{5}{2}$). Particles with a half-integer spin are referred to as the *fermions*, while those with an integer spin as the *bosons*. In the framework of quantum mechanics, the difference in the behavior of fermions and bosons is expressed by the kind of symmetry of the wave functions describing these particles. Without dwelling on the formal foundations of this theory, we only formulate the main conclusion: a system consisting of fermions obeys *Pauli's exclusion principle*, as distinct from a boson system on which no such exclusion principle is imposed. It should be remembered that Pauli's principle reads as follows: *no two fermions may be in exactly the same state.*

An excellent, and probably most important, illustration of Pauli's principle is the atomic level structure underlying the periodic system of the elements. It is known, for example, that the first period of this system is composed of two elements, hydrogen and helium. For the first period, the principal quantum number equals unity. The atomic states associated with the first period are therefore determined only by the value of the spin projection of orbital electrons. As discussed in the preceding section, there are two such values; thus, only two elements can occur in the first period. For the second period, the principal quantum number equals two, giving rise to eight possible different states and thus to eight elements, etc.

It should be emphasized that the Pauli principle is one of the foundations of the very structure of the periodic system. If this principle did not work, all the atomic electrons would populate the ground energy level (i.e., the hydrogen level), and, consequently, the periodicity of the system as well as the valency of chemical bonding would vanish. It is the Pauli principle that prevents atomic electrons from occupying the energetically most favorable ground state.

Another basis for the classification of the elementary particles is their interaction (cf. Table 1.1). All particles participating in the strong interaction are referred to as *hadrons*. All fermions which do not participate in the strong interaction are called *leptons*. Finally, a special place in this classification is reserved for the bosons, particles which mediate the interactions (cf. the next section). The hadrons are well represented by the nucleons and the pions; the leptons, by the electrons and the muons. A typical mediator of interactions is the photon. The hadrons, in turn, are subdivided into the *baryons* and the *mesons*. The baryons are fermions; the lightest baryon is the proton. The hadrons with integer spin are referred to as mesons; the lightest meson is the pion ($m_\pi \sim 140\,\mathrm{MeV}$).

Table 1.1 Classification of the elementary particles

Particle name	Spin	Interactions involving the particles	Mass
Fermions	Half-integer		
		Definition is independent of the interaction and mass of particles	
Bosons	Integer		
Hadrons (fermions or bosons)	Definition is spin-independent	Strong, electromagnetic, weak, gravitational	Definition is independent of particle mass
Leptons (fermions)	Half-integer	Electromagnetic, weak, gravitational	Definition is independent of particle mass
Mesons (a variety of hadrons)	Integer	Strong, electromagnetic, weak, gravitational	Definition is independent of particle mass
Baryons (a variety of hadrons)	Half-integer	Strong, electromagnetic, weak, gravitational	$m \geq m_\mathrm{p}$

1.4 How Elementary Particles Interact

As already mentioned in Sect. 1.1, particles interact by exchange (cf. Fig. 1.1). The exchange in the process of interaction involves not only energy, momentum, and mass, but also the internal quantum numbers: spin, isospin, charge, and color.

The properties of the exchange particles in the context of quantum field theory determine the interaction to a great extent. In particular, by specifying the properties of the photon (i.e., by setting all quantum numbers equal to zero, with the exception of the spin, $s = 1$) the equations of electrodynamics are obtained.

What do the exchange particles have in common? All of them are bosons. The properties of the exchange particles are summarized in Table 1.2.

The *graviton*, a gravitation field quantum, has not been detected owing to its extremely weak interaction. Although most physicists have no doubts about the existence of gravitons with the properties described, some caution is advisable, since the quantum theory of gravitation itself is far from complete. Unfortunately, because of the mentioned weakness of the gravitational field, there is no hope for rapid progress in detecting and investigating gravitons. Although the intermediate $W^{\pm}$- and Z^0-bosons are also hard to detect, their status is much more definite. There is experimental

Table 1.2 Properties of the exchange particles

Interaction	Exchange particle	Mass GeV	Spin of the exchange particle	Electric charge [e]	Iso-spin	Color
Gravitational	Graviton	0	2	0	0	No
Weak	Intermediate bosons[*]:					
	$W^{\pm}$	80	1	± 1	1	No
	Z^0	90	1	0	1	No
Electromagnetic	Photon	0	1	0	0	No
Strong	Gluon	0	1	0	0	Yes

[*] The intermediate bosons form an isospin triplet.

evidence (primarily regarding the neutrino interaction with matter) which is in excellent accord with the intermediate boson properties.

The difficulty of direct experimental confirmation of the existence of the intermediate bosons is a consequence of their large mass. It was only as late as 1981 that a colliding-beam accelerator with sufficient energy for $W^{\pm}$- and Z^0-boson generation started operating at CERN in Switzerland, and by mid-1983, the existence of the $W^{\pm}$- and Z^0-bosons was reliably established.

The gluons, like the quarks, are not observable in the free state. However, in the late 1970s, considerable progress in the indirect verification of gluons was achieved by investigating the annihilation of high-energy positrons and electrons with hadron generation. It turned out that three hadronic jets occur in such processes; two jets are attributed to quarks and the third one, to gluons. The experimental data on the three-jet processes accompanying positron-electron annihilation are in good agreement with the predictions of quantum chromodynamics, which indirectly confirms the existence of gluons, one of the basic elements of that theory.

Table 1.3 Properties of the four kinds of interaction

Interaction	Coupling constant		Interaction radius cm
	Analytic expression	numerical value at $m = m_p$	
Gravitational	$Gm^2/\hbar c$	0.6×10^{-38}	∞
Weak	$g_F m^2 c/\hbar^3$	10^{-5}	10^{-17}
Electromagnetic	$e^2/\hbar c$	$1/137$	∞
Strong	$a/[\ln (m/m_p)]$ $m \gg m_p$	≈ 1	10^{-13}

After these considerations, it would be helpful to list the properties of the four interactions in the form of a table. The data presented in Table 1.3 are actually a summary of the above discussion. A new entry here is the value of the interaction radius. For the gravitational, the weak, and the electromagnetic interaction, the magnitude of the radius r is determined from the uncertainty relation, $r = \hbar/m_B c$, m_B denoting the mass of an exchange particle. In the case of the strong interaction, the interaction radius, r_N, may be regarded either as an empirical constant or as the distance at which the value of the coupling constant α_s becomes unity.[9]

1.5 Unified Field Theories

It appears indisputable that the ultimate goal of particle physics and quantum field theory is to construct a unified theory of all interactions. Estimating the state-of-the-art, one may say that we are at the beginning of a tunnel, but sparkles of dazzling sunshine can be seen at the other end. Einstein's dream, the unification of interactions, is coming true, but on a basis quite different from his philosophy, viz., on the basis of quantum field theory.

Today, this statement might sound trivial to the reader. It seems so natural to use quantum theory that anything else seems absolutely unjustified. The actual history of progress in fundamental physics is not a well-marked highway, however. For decades, till the late 1960s, it seemed that quantum field theory was incapable of overcoming the difficulties which arise in describing even one particular interaction, especially the strong one. Progress has been made by the construction of a theory unifying the electromagnetic and the weak interactions, the development of the quark model, and the formulation of quantum chromodynamics.

Quantum chromodynamics is one of the variants of field theory, and therefore it appears natural and even obvious now that the interactions should be unified within the framework of quantum field theory. As a matter of fact, this theory provides a set of equations which describe all existing interactions and elementary particles in a unique way.

This statement contains an ill-defined notion: "in a unique way"; let us clarify it. The equations of a unified theory have to describe the common properties of the interactions and of the particles, also stressing, however, their differences. Certainly, one may get the impression that one obscure notion ("in a unique way") is substituted by another ("common proper-

---

[9] The strong interaction is caused by the exchange of colored particles.

ties"). This requires further explanation. The latter notion has two aspects: (i) the unified interaction must be describable by a universal coupling constant or set of coupling constants; and (ii) the interaction must correspond to a common symmetry type. It is this common symmetry which characterizes the common nature of the properties of a set of various particles. Mathematically, this common symmetry corresponds to a common group of transformations under which the unified field theory equations are invariant. Let us consider both aspects of the unified theory in succession.

1.5.1 The Universal Constant

At first glance, it might appear absurd to even question the universality of the coupling constant for all interactions. This is not so, however, if one remembers that the coupling constants α are running constants, i.e., they depend on the mass m (the momentum) of the exchange particles transferred (cf. Table 1.3).

This m-dependence of the coupling constants α may be more or less pronounced, but it is very weak for α_e, the coupling constant of the electromagnetic interaction.[10] In the following, we shall neglect this dependence, assuming $\alpha_e(m) = \text{const}\ (m)$. From a comparison of the different constants listed in Table 1.3, the values of the characteristic masses m_{we}, m_{wes}, and m_{wesg} are then obtained which correspond to the unification of the coupling constants. These quantities are given in Table 1.4.

Table 1.4 Values of the characteristic masses m_{we}, m_{wes}, and m_{wesg}

Interactions to be unified	Symbol of the mass of the interaction mediator	Numerical value of the mass corresponding to unification GeV
Weak, electromagnetic	m_{we}	10^2
Weak, electromagnetic, strong	m_{wes}	10^{15}
Weak, electromagnetic, strong, gravitational	m_{wesg}	10^{19}

The unification of all four interactions is the least advanced. Since we shall not come back to the gravitational interaction later on, let us just note that the mass m_{wesg} is usually set to be

---

[10] This dependence has not been mentioned so far.

$$m_{\text{wesg}} = \left(\frac{\hbar c}{G}\right)^{1/2} = \alpha_g^{-1/2} m_p \quad .$$

The mass corresponding to such a unification is determined through fundamental constants only, and was defined by Planck early in this century. In the opinion of many physicists, it should play a major role in the unification of all interactions.

But let us now return to our main subject, the unification of the remaining three interactions. Since the lowest value of the mass associated with unification corresponds to the combination of the weak and the electromagnetic interactions ($m_{\text{we}} \sim 100\,\text{GeV}$), it is instructive to begin by considering their unification.

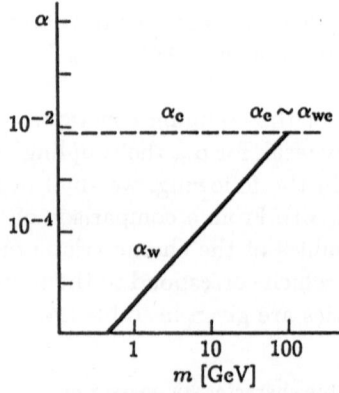

Fig. 1.2. Dependence of the constants α_w and α_e on the transmitted mass m (schematic). A weak dependence of the constant α_e on m has been neglected in this plot

Under the above assumption (α_e=const), only one constant, $\alpha_{\text{we}} \sim \alpha_e$, arises in the range $m > m_{\text{we}} \sim 100\,\text{GeV}$, where the dependence $\alpha_{\text{we}}(m)$ is not strongly pronounced. The m-dependence of the constants α_e and α_w is shown in Fig. 1.2. It should be noted, however, that as a matter of fact, the unified electroweak interaction is characterized by two common constants and two masses (cf. Table 1.2) which differ slightly and merge at a higher mass range. This feature is caused by the specific structure of the field mediating the weak interaction (see below). The behavior of the two constants, α_{we}^1 and α_{we}^2, as well as of the constant α_s with m is shown in Figs. 1.3. It is quite symptomatic that all three running constants become equal around $m_{\text{wes}} \sim 10^{15}\,\text{GeV}$. This circumstance does not seem to be accidental: it seriously substantiates the notion of the unification of three interactions — strong, electromagnetic, and weak — which is referred to as the *grand uni-*

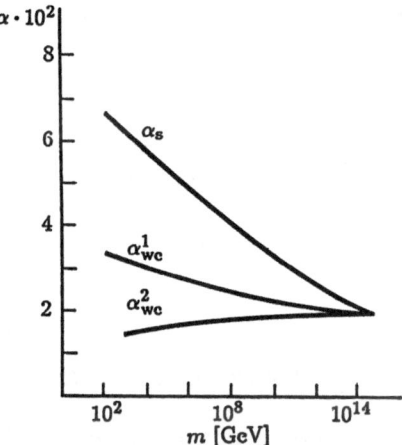

Fig. 1.3. Dependence of the constants of the unified electroweak interaction and of the constant α_s on m

fication. The value of m_{wes} corresponding to the unification of the three constants can be approximately obtained from the following considerations (cf. Table 1.3). Setting $\alpha_s = \alpha_e$ one gets

$$m_{\text{wes}} \sim m_{\text{p}} \exp\left(a/\alpha_e\right) \qquad (1.4)$$

where $a \sim 1$ depends on the number N_q of the kinds of quarks. A more precise evaluation of a can be obtained by using the exact expression relating the constant α_s to the mass m_{wes}. For $N_q = 6$ it reads

$$\ln \frac{m_{\text{wes}}}{m} = \frac{\pi}{11}\left[\frac{1}{\alpha_e} - \frac{8}{3\alpha_s(m)}\right] \qquad . \qquad (1.5)$$

It follows from experiment that $\alpha_s(m_{\text{p}}) \sim 1$. Hence, by setting $m = m_{\text{p}}$ and $\alpha_s(m_{\text{p}}) \sim 1$ and taking into account that $1/\alpha_e$ is much larger than $8/[3\alpha_s(m)]$ (note that $\alpha_e = 1/137$), (1.5) can be approximately expressed as

$$\ln \left(\frac{m_{\text{wes}}}{m_{\text{p}}}\right) \sim \frac{1}{(4\alpha_e)} \qquad . \qquad (1.6)$$

It follows eventually that

$$m_{\text{wes}} = m_{\text{p}} \exp\left[1/(4\alpha_e)\right] \sim 10^{15}\,\text{GeV} \qquad . \qquad (1.7)$$

17

The mass scale arising within the grand unification is enormous. It should be noted that at present particles with mass $\sim 100\,\text{GeV}$ are accessible to study by colliding-beam accelerators. Perhaps more impressive is the fact that even if such a fantastically difficult feat as the construction of a circular accelerator around the Earth were possible, particles with an energy of only $\sim 10^7\,\text{GeV}$ could be generated, i.e., many orders of magnitude smaller than the mass m_{wes}. Nevertheless, as will be seen below, there is an experimental approach to investigating certain characteristics of particles with the mass m_{wes} (the proton decay).

1.5.2 Unified Symmetry

The statement that a certain symmetry type underlies the dynamic equations appears trivial nowadays. The idea of unification is based on two postulates: (i) There exists a symmetry type common to the interactions; and (ii) there exists a symmetry type common to and characteristic of the fundamental elementary particles.[11]

Let us begin with the first question. Earlier, examples of symmetry, geometrical and the internal, were given. However, geometrical symmetries determine only the laws of conservation of energy, momentum, and angular momentum, i.e., general laws valid for any dynamic equations. In this sense they in no way promote the search for equations of a unified theory. Much more restrictive is the role of the internal quantum numbers. It has been mentioned already that the quantum numbers of the exchange particles determine the character of an interaction. In this approach, however, the interactions appear isolated, because the quantum numbers of the exchange particles seem to be quite different. This is true on one stipulation: the spins of the exchange particles are equal for three interactions, their value being unity (cf. Table 1.2). It is this circumstance which provides these interactions with a common ground that facilitates unification. The gravitation spin equals 2, making inclusion of the gravitational interaction in a unified theory difficult.

Let us come back, however, to analyzing a possible formulation of the general dynamic principle. Currently, special hope is held for the *principle of gauge invariance*, which is obeyed by all four interactions.

Let us consider this principle in some detail. A simple illustration of it are the high-school experiments on electromagnetism. One can touch one of the wires through which a current flows with one's bare hand, but every reader has probably had the unpleasant experience of touching both

---

[11] The notion of fundamental particles will be explained below.

wires simultaneously. As is well known, this simple empirical fact can be explained in an elementary way: The current is determined by the potential difference, and not by the potential itself. By adding a constant to the value of the potential (or by subtracting one), no change in the physical picture is produced. In other words, there are quantities in the theory, viz., potentials, whose absolute value does not affect the measured parameters. These are determined by the potential difference, and the theory is thus invariant with respect to the parameter change by a constant. In electrodynamics, the vector potential, consisting of four components which transform as the components of a four-vector is a generalization of high-school considerations of the potential. This property of the independence of physical effects of even more general transformations (known as gradient transformations) is referred to in electrodynamics as gauge invariance.

Another example is borrowed from radio engineering. The radio wave interference depends on the phase shift rather than on the absolute values of the phases.

The above examples are particular cases of the gauge principle which can be formulated as the invariance of physically measurable quantities with respect to transformations of other quantities (potentials) which are not measured experimentally. Potentials and measured quantities, e.g., forces, are connected by simple transformations. Detailed investigations have demonstrated the universality of gauge invariance. Furthermore, by specifying the form of gauge invariance as well as the parameters of the particles mediating an interaction, all properties of this interaction can be determined. The universality of gauge invariance is one of the cornerstones of interaction unification.

The other one is connected with a generalization of the properties of the fundamental elementary particles. The very combination of the attributes "fundamental" and "elementary" is not well defined. One can nevertheless try to describe this category in brief, though. The *fundamental elementary particles* are pointlike fermions, i.e., particles whose spatial dimensions are zero. Of course, the question of the physical foundations of this definition arises. Fundamental particles are understood as the objects of which the rest of the elementary particles are made up. Clearly, such particles must be fermions. Indeed, from particles with a half-integer spin, bosons can be formed which possess an integer spin ($\frac{1}{2} + \frac{1}{2} = 1$); the opposite is not true: Fermions cannot be made up of bosons ($1 + 0 \neq \frac{1}{2}$). Extended particles cannot be fundamental (i.e., truly elementary).

It turns out that of the large variety of particles only very few actually fall in the category of fundamental particles as defined here. First to be mentioned are the leptons (whose number currently amounts to five) and, possibly, the quarks. That the leptons can be qualified as truly elementary follows from theoretical considerations and from the experimental data indicating that the size of the electrons and the muons is smaller than 10^{-15} cm. Quantum electrodynamics, a theory which describes the behavior of charged leptons, is based on the concept that they are pointlike, and it is in excellent agreement with experiment.

The assumption of the fundamentality of quarks is more problematic. Certainly, it is not possible to reliably measure quark size; Experiments, however, are not in contradiction with their pointlikeness. The fact that all the numerous hadrons consist of quarks is a serious argument in favor of their "fundamentality."

Thus, it is postulated that unification ought to be performed on the basis of leptons and quarks. The electroweak interaction does not include the strong one; hence, it can only involve leptons. The simplest version of unification is to assume that each lepton and its associated neutrino form an analogue of the isospin family,

$$\begin{pmatrix} e \\ \nu_e \end{pmatrix}; \quad \begin{pmatrix} \mu \\ \nu_\mu \end{pmatrix}; \quad \begin{pmatrix} \tau \\ \nu_\tau \end{pmatrix} \quad .$$

Here, ν_τ denotes the neutrino corresponding to the τ-lepton. This neutrino has not been observed as yet, but there is no doubt concerning its existence.

The unification of the exchange particles is more complicated. In the framework of the electroweak interaction, one has to unite the photon — mediator of the electromagnetic interation — and the heavy bosons (with mass $m_{we} \sim 100$ GeV) — mediators of the weak interaction. Two competing schemes for unifying the mediating particles into one family were proposed: According to the first scheme, the mediation of the electroweak interaction occurs by a photon and an isospin triplet of $(W^\pm, Z^0)$-bosons; in the second one the mediators of the electroweak interaction form an isospin triplet consisting of two charged $W^\pm$-bosons and a photon as the third component. Experiment has demonstrated the validity of the first version which earned its authors (Glashow, Salam and Weinberg) a Nobel Prize in 1979.

Hence, the electroweak interaction is currently understood as follows: For small energies, the electromagnetic interaction is mediated by the photon, while the weak interaction is mediated by the $W^\pm$- and Z^0-bosons; for large transferred masses $m \gtrsim m_{we}$, a unified interaction exists, mediated by all four particles.

The grand unification includes, in addition, the strong interaction; that is to say, the leptons and quarks are combined into one group. The quark-lepton symmetry implies that the quarks and leptons occur in equal numbers. At present, it is usually thought that the number N_q of quark types is equal to six; so far, there are serious arguments for the existence of five sorts of quarks. It is believed that a sixth quark has a very large mass and may be detected with the help of new accelerators.

Let us proceed to the particles mediating the strong interaction. It turns out that even in the simplest versions of the grand unification, the four particles mediating the electroweak interaction and the eight gluons mediating the strong interaction have to be complemented by 12 superheavy X-particles with mass $m_{wes} \sim 10^{15}$ GeV, which simultaneously mediate the weak, the electromagnetic, and the strong interactions.

1.6 Proton Decay

The proton is a very stable particle. Its lifetime t_p definitely exceeds 10^{15} years – a figure five orders of magnitude larger than the age of the Universe, 10^{10} years. The evidence for this obvious fact is our very existence. A human body contains $\sim 10^{29}$ protons. If the time t_p were smaller than 10^{15} years, more than 10^{14} protons would decay per year. Ionization caused by these decays would suffice to exterminate all the sizeable creatures, including all of mankind, of course. The time $t_p \sim 10^{15}$ years is a tremendous length of time, even compared with the age of the Universe. It was therefore quite natural to consider this time as infinite, i.e., to regard the proton as an absolutely stable particle. Almost all physicists grew accustomed to this concept, and only now and then sceptics raised the question: Why doesn't the proton decay? The question stemmed from the following dilemma: On the one hand, public opinion is based on experience, and experience claimed the stability of the proton; on the other hand, the entire bulk of experimental material (not related to the proton stability) supported the statement that everything should occur in Nature if not prohibited (cf. Sect. 1.2.3).

In the Universe, there are two stable particles with nonzero mass, the electron and the proton. It is clear why the electron does not decay. The electron is the lightest of the charged particles; all lighter particles (photon, neutrino) are electrically neutral. So it is the charge conservation law which prevents the electron from decaying. Why the proton should not decay is unclear, although there are many potential channels possible (e.g., into a pion and a neutrino). To reconcile this fact with the rule that "everything admissible in the world of elementary particles does occur," the *law of conservation of baryonic charge* was devised post factum, the baryon number $+1$ and -1 being attributed to all baryons (including the proton) and antibaryons, respectively. This law, then, provided a simple "explanation" of proton stability. The proton charge is $+1$, while that of all lighter particles is zero. For example, in the reaction $p \rightarrow \pi^+ + \nu$, the baryonic charge on the left-hand side equals +1, whereas that on the right-hand side is zero. For this reason, proton decay proves to be impossible. This simple rule formally resolved all doubts, but there were still some muffled sceptical voices to be heard.

The law of conservation of baryonic charge was introduced by analogy with the law of conservation of electric charge. In addition to being a conserved quantity, the electric charge has another extremely important function: it is a quantitative measure of the electromagnetic interaction. The baryonic charge, on the contrary, has not such function. It was possible to demonstrate experimentally with a high degree of accuracy that the baryonic charge is not involved in long-range interactions. The credibility of the analogy between the baryonic charge and electric charges was undermined by this fact. As a result, doubts about the absolute stability of the proton were expressed from time to time, motivating specially devised experiments to check on it; thus it was found in the early 1970s that $t_p \geq 10^{29}$ years. Soon, the problem of the stability of the proton became a major concern of elementary particle physics.

This radical change of viewpoints concerning the stability of the proton resulted from the successes in unifying the interactions, notably from the grand unification. At the end of the preceding section dealing with this question it was stated that in the framework of the grand unification, exchange X-particles exist which have a fractional charge and which are capable of simultaneously mediating the weak, the electromagnetic, and the strong interactions. A proton consists of three quarks (Fig. 1.4); an X-particle leav-

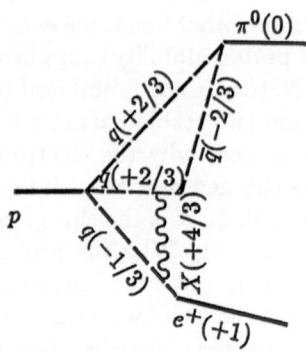

Fig. 1.4. Schematic representation of proton decay. The proton first decays into three quarks with the electric charges $\frac{2}{3}e$, $\frac{2}{3}e$, and $-\frac{1}{3}e$. Two quarks then interact with each other via exchange of a heavy boson with the charge $\frac{4}{3}e$. In the process of this exchange one of the quarks acquires an integer charge and transforms into a positron. The quark with the charge $\frac{2}{3}e$ transforms into an antiquark (with the charge $-\frac{2}{3}e$) to form a pion by associating itself with the other quark of charge $\frac{2}{3}e$

ing one of the quarks carries away its characteristic quantum numbers — the electric charge and color — thus permitting its transformation into a lepton; the two remaining quarks form a light hadron, e.g., a pion. This is an outline of how instability of the proton is conceived of within the grand unification.

At first glance this unexpected conclusion seems to contradict the experimental data on the stability of the proton. However, the greatest intriguing significance of the prediction of the grand unification is that the proton lifetime $t_p = 10^{31\pm2}$ years turned out to be close to the experimental limit of t_p and, at the same time, the theoretical value of t_p is within reach of modern experimental techniques. No wonder that the theoretical prediction following from the grand unification theory was a sensation for the scientific, but not only the scientific, community; the discussion of proton decay was even taken up by the mass media.

Now, the most important question arises: Where does the enormous time scale of t_p come from? A possible answer is that the time t_p is associated with the large value of the mass of the X-boson, $m_{wes} \sim 10^{15}$ GeV. In fact, the decay probability, W_d, is proportional to the product of the emission and absorption cross sections for an X-boson. A cross section has the dimension of the square of the length. For large masses m_{wes}, there is only one characteristic length, i.e., the Compton length for the X-boson, $\hbar/m_{wes}c$; hence, $W_d \propto (m_{wes})^{-4}$. The lifetime is inversely proportional to W_d, i.e., $t_p \propto (m_{wes})^4$.

This is a qualitative argument. A more precise expression can be obtained by taking into account the fact that the time t_p is approximately proportional to α_e^{-2} and to the time $\hbar/m_p c^2 \sim 10^{-24}$ s. Then one has

$$t_p \sim A\alpha_e^{-2}(m_{wes})^4 \left(\frac{\hbar}{m_p c^2} \right) \tag{1.8}$$

From dimensionality considerations it follows that the dimension $[A] = [m^{-4}]$. In the world of elementary particles, the characteristic mass is the proton mass. By setting $m = m_p$, $A \sim 1/m_p^4$, and $m_{wes} = m_p \exp\left[1/(4\alpha_e)\right]$ (cf. (1.7)), an elegant approximation for the time t_p is obtained

$$t_p \sim \alpha_e^{-2} \exp\left(\frac{1}{\alpha_e} \right)\left(\frac{\hbar}{m_p c^2} \right) \sim 10^{32} \text{ years} \tag{1.9}$$

The large value of t_p is primarily determined by the smallness of the coupling constant α_e as compared to unity.

Hence, the grand unification leads to a completely unexpected conclusion: the proton must decay; its lifetime t_p will be very large, however. This conclusion has principal implications in many respects: (i) detecting this decay would confirm the validity of the grand unification principle; (ii) establishing decay would prove the validity of the equations of chromodynamics as well as their applicability to the quarks as truly elementary particles; and (iii) a more detailed investigation of proton decay would provide an insight into the world of particles of the fantastically large mass $\sim 10^{15}$ GeV.

It is no wonder that many large groups of physicists are involved in searching for proton decay. By the second half of 1982, there were 11 groups, representing almost all the continents, actively searching for, or planning to search for, decaying protons. Seven experiments were in 1982 in operation, with proton-decay targets weighing about 100 tons. The first to start this research were the Soviet physicists (the Baksan Neutrino Observatory) as well as a joint Japanese-Indian group (several Japanese universities and the Tata Institute of Fundamental Research), who works in a goldmine in India. According to the experimental data, $t_p \sim 10^{31}$ years. However, the results obtained by the Japanese-Indian group admit dual interpretation. According to their data, several cases were recorded which can be interpreted as proton decays. If this interpretation is correct, then the Japanese-Indian group has measured the value of the time t_p, namely, $t_p \sim 5 \times 10^{30}$ years, rather than its limit.

It might appear that the moment has come to be more optimistic and to publicize the discovery of proton decay. Unfortunately, the situation is not so simple. The cases which have been interpreted as proton decays occurred within the target but in the proximity of its surface, so that boundary effects causing inaccuracy of interpretation cannot be ruled out.[12] A discussion kindled by these few proton decay candidates shows that the question of decay has not been finally solved, and that the suspense is mounting; a proton decay may be detected any moment.

A natural question is why such a large number of experimental groups cannot succeed in resolving the clearly stated problem of whether or not protons decay; it is the large magnitude of the time t_p which causes the difficulty. Supposing t_p is about 10^{30} years; then one decay per annum will take place within one ton of matter. For $t_p \sim 10^{32}$ years, one decay per annum will occur within 100 tons of matter. Consequently, these extremely rare events can be reliably recorded only in large amounts of matter. Solving this problem would only mean partial success, however. As a matter of fact, in any experimental device there always will be a background, more or less pronounced, that is generated by radioactive impurities or by cosmic

---

[12] According to the data reported by the Japanese-Indian group at the Neutrino-82 Conference held in Budapest in June 1982, three events have been recorded far from the outer surface of the unit which cannot be accounted for by background effects. If these events are to be associated with proton decays, then $t_p \sim 6 \times 10^{30}$ years. However, there are some doubts as to whether these events, too, are interpretable as proton decays.

24

rays and that simulates proton decay events. One cannot get rid of this background completely, but it can be significantly reduced. To reduce the background originating from cosmic rays, the detectors have to be set up under ground, the cosmic radiation being absorbed by the depths of earth. For example, the Japanese-Indian group mentioned above works at a depth of 3 km which enables them to reduce the cosmic background by about 7–8 orders of magnitude. To suppress the radioactive background, attempts are being made to use extremely purified substances free of radioactive impurities; thorough purification of a large bulk of material is a complex technological problem per se, however.

What are the prospects for studying proton decay in view of the difficulties discussed? One of the main directions in designing experimental setups is to use purified water as the major element of the target and to employ Čerenkov radiation counters as decay detectors. The parameters of the detectors are as follows: the useful weight of the target (water) is about 10 000 tons; the number of big Čerenkov counters amounts to several thousands. One such experiment is in operation in the USA (Ohio).[13]

Experimental set-ups of this kind make it possible to detect proton decay, provided that $t_p < 10^{33}$ years, which is close to the principal detectability limit of decay. Indeed, if t_p proves to be $\sim 10^{34}$ years, the background stemming from cosmic neutrinos, which cannot be suppressed, will inevitably imitate proton decays. It remains to be hoped that $t_p < 10^{34}$ years and that proton decay will be found in the not-so-distant future. This would lead to a radical advance in the physics of elementary particles.

This is how the prospects for studying proton decay gleam through the haze of fundamental physical hypotheses and methodological problems.

---

[13] According to the experimental data obtained with this detector $t_p > 10^{32}$ years, in contradiction to the simplest (minimal) model of Grand Unification which predicts $t_p \sim 3 \times 10^{31}$ years.

2. The Universe

2.1 A Bit of History

Ever since Man grew interested in the structure of the macrocosm, the belief prevailed that the stars are an eternal and never changing ornament decorating the night sky. Of course, it did not remain unnoticed that in the background of this impressive permanency on a huge scale, there are several stars (planets) whose positions with respect to the Earth do vary. A consistent system arose, synthesizing a tremendous amount of empirical material accumulated by ancient oriental and Greek astronomers. Their world consisted of immovable stars "pinned up" to a rotating sphere, as well as of the planets and the Sun moving around the Earth, the center of the Universe.

The revolutionary views of Copernicus and Galileo – as we know today – only touched upon a tiny part of the Universe – the solar system – and replaced the Earth as the "center of the Universe" by the Sun. As a matter of fact, the view of an invariable firmament persisted till the beginning of this century.

In contrast to other disciplines where evolutionary ideas were put forward and settled long ago (cf., e.g., Cuvier's theory of cataclysms or Darwin's theory of the origin of the species), the oldest of the sciences, astronomy, was dominated by the belief that the world is endless and invariable. Probably the respectable history of this belief played a decisive role in Einstein's attempts to apply his theory of general relativity to describe the Universe, in particular, to explain its invariablity. Since the invariablity of the Universe did not directly follow from Einstein's equations, he included in them a so-called λ-term incorporating hypothetical forces to impede the Universe from flying asunder. In this way, Einstein was successful in constructing a model of a Steady-State Universe (1917).

Soon hereafter, immense changes occurred in the history of mankind, and an unobtrusive event remained unnoticed: in the early 1920s, in Petrograd* a group of young enthusiastic neophytes started studying the theory

---

* (Former name (1914–1924) of Leningrad)

of general relativity. Among them was A.A. Friedmann, a mathematician and meteorologist. Friedmann succeeded in solving the equations of the theory of general relativity (without the λ-term) under quite general and natural assumptions. He demonstrated that the theory leads to an unexpected result: The Universe must be nonstationary (cf. Sects. 2.2 and 4.5). After some discussion, Einstein accepted this result, as reflected in his book entitled *The Meaning of Relativity*.[1]

However, acknowledgement by Einstein did not mean acknowledgement by the scientific community, which considered Friedmann's result to be mathematically curious rather than of any profound physical significance. It was only in 1929, after E. Hubble's observation of the effect of galactic redshift, that an almost instantaneous change in public opinion took place: The Universe was now regarded as non-steady-state. Friedmann did not live to see the triumph of his prediction; he died in 1925 at the age of 37.

It would not be exaggerated to say that the achievements of Friedmann and Hubble were a turning-point in the history of astronomy. They put an end to a major fallacious notion; that of the invariability of the Universe. This result has far-reaching consequences. The recognition of the evolution of the Universe in space-time has another implication for the world outlook, namely, that such cosmic objects as galaxies and stars evolve as well. The recognition of the variability of these physical objects required some time, of course. Only some two decades after the works by Friedmann and Hubble did the notion of the invariability of the Universe yield to the idea of universal evolution. It is doubtless, however, that the conclusions made by Friedmann and Hubble were one of the two factors which stimulated a major change in the picture of the Universe. The other factor was the technological progress that made it possible to perform astronomical observations at almost all frequencies of the electromagnetic spectrum.

This progress involved three directions: the development of electronics, aeronautical engineering, and instrumentation. Radio astronomy is a by-product of the invention of radar which makes it possible to observe the cosmos in the radio-frequency range. But only in the radio range ($E_\gamma < 10^{-3}$ eV)[2] and in the optical range ($E_\gamma \sim 1 - 3$ eV) is radiation transmitted through the atmosphere almost without absorption.

In the infrared and sub-millimeter ($E_\gamma \sim 10^{-3} - 1$ eV), ultraviolet ($E_\gamma \sim 3$–100 eV), X-ray ($E_\gamma \sim 100 - 10^6$ eV), and gamma-ray ($E_\gamma > 10^6$ eV) ranges, radiation is absorbed by the atmosphere, and detectors for this radiation have to be placed at the boundary of the atmosphere or beyond

---

[1]   A. Einstein: *The Meaning of Relativity* (Princeton, N.J., 1953).

[2]   The energy of quanta E_γ (as well as the frequency of radiation) are measured in eV. A quantum with the energy of 1 eV has the wave length $\lambda \sim 10^4$ Å $= 10^{-4}$ cm.

it. For this purpose, balloons, rockets, and satellites are used. Of course, recording devices have nothing to do with optical telescopes; such devices combine the advances of microelectronics and nuclear physics. For example, in gamma-ray astronomy, spark chambers with very sensitive track registration are used as a basic element of the recording devices. Without going into detail, let us just mention that for each frequency range, there is a specific technological solution.

Friedmann's ideas and Hubble's observations as well as technological progress led to an essential change in the very object of astronomy. Until the middle of the 1940s, mainly quasi-steady-state objects were investigated, their lifetime being equal to the age of the Universe. After the fifties, the priorities were displaced toward non-steady-state or explosive objects with very different time scales. For example, the duration of gamma-ray bursts[3] is of the order of 10 s, while the duration of the activity phase of galactic nuclei and quasars is of the order of 10^6 to 10^7 years.

The next question to be asked is certainly: What is the impact of technological progress and, as a consequence, of a tremendous broadening of the observable frequency range on proving the evolution or explosions of cosmic objects? The answer is that the energy spectrum of radiation from quasi-steady-state objects has an equilibrium character and is, as a rule, well represented by Planck's distribution with a sharp maximum in the region of $E \sim 1\,\mathrm{eV}$, which corresponds to the optical range; cf. Fig. 2.1a. The energy spectrum of non-steady-state objects is essentially nonequilibrium; it does not exhibit a maximum and contains radiation corresponding to all

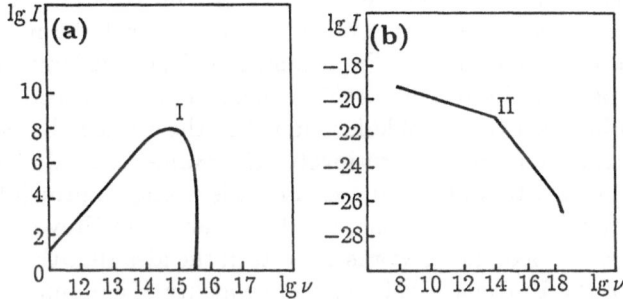

Fig. 2.1a,b. Intensity of radiation of the Sun (a quasi-steady-state object) (a) and of the Crab Nebula (a non-steady-state object) (b). The solar radiation spectrum is given by the Planck distribution, ν being the frequency of radiation

---

[3] Gamma-ray bursts are short pulses of γ-radiation of cosmic origin. According to a currently popular hypothesis, the source of γ-bursts are neutron stars. The particular mechanism of the instability of neutron stars giving rise to the bursts is not so clear, however.

ranges, from the radio frequency to the gamma-ray frequency range. (An example of such a spectrum is depicted in Fig. 2.1b.) Investigating non-steady-state objects is only possible with the aid of the observations in all frequency-bands.

2.2 Friedmann's Model of the Universe

The model of the Universe proposed by Friedmann is based on two postulates referred to as the Cosmological Principle (cf. Sect. 4.5): (i) the Universe is isotropic in three-dimensional space; and (ii), the Universe is homogeneous in three-dimensional space. From this principle, an extremely important conclusion follows, irrespective of the evolutionary dynamics of the Universe. Let us consider two point-like bodies a distance r_{12} apart. The Universe being isotropic, the direction of the vector r_{12} is the only preferred orientation in the Universe. Therefore, all vectorial dynamic quantities determining the state of this two-body system have to be directed along the vector r_{12}. In particular, this refers to the relative velocity. Hence, the relation holds:

$$v_{12} = H r_{12} \quad . \tag{2.1}$$

Generally, the parameter H should depend on the position of points 1 and 2. But the second cosmological postulate, i.e., that of the homogeneity of the Universe (ii), suggests that $H = \text{const}(r)$; that is to say, at a given moment of time, t, the quantity H is a constant, universal for the entire Universe. This constant is referred to as the Hubble constant. As mentioned in Sect. 2.1, relation (2.1) was revealed by Hubble in 1929: it was the first experimental substantiation of Friedmann's model of the Universe based on the theory of general relativity. At first glance, our consideration belies any connection between the Hubble law and that theory: it is based on simple reflexions and does not seem to involve the evolutionary dynamics of the Universe. As a matter of fact, this argument is wrong: in principle, Hubble's relation is fulfilled for $H = 0$, i.e., for the steady-state Universe, as well. A lasting value of Friedmann's works consists in the idea that the evolutionary dynamics of the Universe results in the inequality $H \neq 0$, implying that the Universe must be a non-steady-state object.

Here, the irony of fate can be recognized: although the first proposed model of an expanding Universe was based on the theory of relativity, its main results can be derived within the Newtonian theory of gravitation. This fact, though somewhat paradoxical from the historical viewpoint, was demonstrated by the British astrophysicists Milne and McCrea in 1934, more than 10 years after the publication of Friedmann's works. The non-steady-state character of the Universe follows from simple, even obvious, considerations. If it were not for the belief in an eternally unchangeable Universe clouding logical thinking and imagination, the "Friedmann model" could have been already discovered by Newton.

Let us make this clear by way of example. Consider a body thrown up from the Earth (a launched rocket, to be specific). Should the velocity of the projectile be smaller than the first cosmic velocity (i.e., the orbital velocity, $\sim 8\,\mathrm{km\,s^{-1}}$), it will fall back to Earth. If its speed exceeds the second cosmic velocity (escape velocity, $\sim 11\,\mathrm{km\,s^{-1}}$), the projectile will never return to Earth, becoming an interplanetary probe. Should the velocity be about $10\,\mathrm{km\,s^{-1}}$, the projectile will become a satellite of the Earth, i.e., it will follow a circular path around the Earth, the center of this circular motion. However, there should be no such center in an isotropic and homogeneous Universe! There is no preferential point in such a Universe about which a body could move on a circular orbit. Consequently, there should be no closed steady-state orbits. The existence of steady-state cosmic systems (planets, galaxies, etc.) does not discard this conclusion, for the Cosmological Principle itself only holds in an approximate way: on a small scale, the Universe is inhomogeneous,[4] so that the Cosmological Principle is to be understood as the requirement of homogeneity and isotropy on a scale comparable with the size of the Universe, $R_\mathrm{u} \sim 10^{28}$ cm.

Hence, under the above assumption, only two outcomes of the evolution of the Universe are possible: one analogous with the return of the projectile, and the other analogous with its escape. Both imply that the Universe is non-steady-state.

2.3 Evolution of the Universe: A Quantitative Analysis

This model of the Universe, based on Newton's theory of gravitation, involves a fundamental assumption which cannot be qualified as a postulate, for it follows from modern experimental data: it is the assumption that the dynamics of the expansion of the Universe is determined exclusively by the gravitational interaction. In fact, for particle spacing r exceeding the hadron size $r_N (r_N \sim 10^{-13}$ cm), the short-range interactions (i.e., the strong and the weak ones) cannot play a role. In the early stages of expansion, for $r < 10^{-13}$ cm, the short-range interactions do not affect the expansion dynamics for a somewhat different reason: these forces act between two or more particles, whereas the action of gravitation on each particle involves the forces from all particles of the Universe. The long-range electromagnetic

---

[4] We shall return to this question in Sect. 2.8.

interaction has no effect because of the electric neutrality of the Universe, the forces caused by the positive and the negative charges cancelling each other.

Thus, the destinies of the Universe as a whole are determined by the gravitational interaction. The postulates of homogeneity and isotropy can be roughly visualized by considering a homogeneous isotropic sphere. Consequently, in the model being discussed (i.e., in the Newtonian approximation), the dynamics of the Universe is reduced to the problem of the evolution of a homogeneous isotropic sphere in its own gravitational field.

To solve this problem, we shall make use of a result of classical Newtonian theory which will be given without mathematical derivation: any point-like body placed anywhere inside a uniform sphere of radius R experiences a gravitational force from the particles from within the sphere of radius r. (Here, r denotes the distance from the body to the center of the sphere, cf. Fig. 2.2). The resultant force exerted on this body by the particles lying outside the sphere of radius r is zero, implying the perfect annulment of individual contributions. The effect of the particles from within the sphere of radius r is the same as if the total mass inside it were concentrated at its center. In other words, the force F acting on a body of mass m at a distance r from the center is given by

$$F = -\frac{GMm}{r^2} \quad, \tag{2.2}$$

where M is the total mass of matter inside the sphere of radius r; the minus sign accounts for the attractive character of the gravitational forces. Henceforth, we shall set $m = 1$, $r = R_u$, and $M = M_u$, where R_u and M_u denote the radius and the mass of the Universe, respectively. Equation (2.2) can then be rewritten as

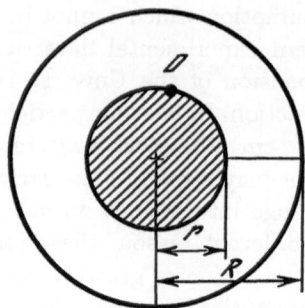

Fig. 2.2. The effect of gravitation upon an element 0 inside a uniform gravitating sphere. The element 0 is affected only by the matter contained within the sphere of radius r.

$$\frac{d^2 R_\mathrm{u}}{dt_\mathrm{u}^2} = -\frac{GM_\mathrm{u}}{R_\mathrm{u}^2} \ .\tag{2.3}$$

Assuming that the mass M_u of the sphere does not vary[5] in the process of the evolution of the Universe, this equation can be easily integrated:

$$\frac{1}{2}\left(\frac{dR_\mathrm{u}}{dt_\mathrm{u}}\right)^2 - \frac{GM_\mathrm{u}}{R_\mathrm{u}} = E \ ,\tag{2.4}$$

E being a constant of integration. The last equation can be given a simple interpretation. Indeed, it could be written down without solving (2.3). The relation (2.4) expresses the principle of conservation of energy as applied to an "isolated" boundary element of the Universe of mass $m = 1$. "Isolation" is justified by Hubble's law: all elements inside the sphere have a velocity smaller than the velocity $V = dR_\mathrm{u}/dt_\mathrm{u}$ of the element lying on the boundary R_u and therefore they cannot catch up with it.

Very important conclusions can be drawn from (2.4). If $E>0$, then the velocity $V = dR_\mathrm{u}/dt_\mathrm{u}$ can never vanish, the term $GM_\mathrm{u}/R_\mathrm{u}$ always being positive; for $E<0$, however, there is a value of R_u for which $V = 0$. This conclusion can be interpreted as follows: the case $E>0$ corresponds to the kinetic energy $V^2/2$ exceeding the potential energy $GM_\mathrm{u}/R_\mathrm{u}$. Accordingly, once started, sphere expansion will go on indefinitely (note the analogy to the escape velocity). This case corresponds to the so-called open Universe. In the opposite case of $E<0$, the equality $V = 0$ is possible. At the moment the velocity vanishes, the expansion of the Universe will be superseded by its contraction. This case is referred to as that of a closed Universe. The dependence $R_\mathrm{u}(t_\mathrm{u})$ for a closed and an open Universe is shown in Fig. 2.3.

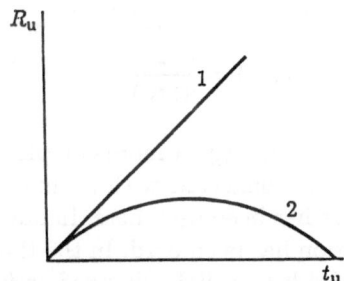

Fig. 2.3. The dependence $R_\mathrm{u}(t_\mathrm{u})$ for an open (*1*) and a closed (*2*) Universe

---

[5] This assumption corresponds to neglecting the effect of radiation, which will be discussed in the next section.

Equation (2.4) can be represented in a more convenient form in which the evolutionary characteristics of the Universe are expressed in a particularly simple way. Using the Hubble law, $dR_u/dt_u = HR_u$, and introducing the average density ϱ through the relation $M_u = \frac{4}{3}\pi\varrho R_u^3$ transforms (2.4) into

$$H^2 - \left(\frac{8}{3}\right)\pi\varrho G = \frac{2E}{R_u^2} \quad . \tag{2.5}$$

It follows from this equation that the Universe is open, i.e., expands without limit ($E>0$), provided that $\varrho<\varrho_c$. The Universe is closed, i.e., the expansion is followed by contraction ($E<0$), for $\varrho>\varrho_c$. The quantity $\varrho_c = 3H^2/8\pi G$ is called the critical density. The character of the evolution of the Universe is determined by this simple relation between ϱ and H. For the current epoch, $H_0 \sim (3 - 5)\,10^{-18}\,\mathrm{s}^{-1}$, the corresponding value of the density is $\varrho_{c0} \sim 10^{-29}\,\mathrm{g\,cm}^{-3}$. (The subscript 0 indicates that the quantities refer to the current epoch.)

Quantitative description of the evolution of the Universe requires solving (2.4) or (2.5). To simplify the procedure, we set $E = 0$. This simplification can be justified by two facts. First, it is supported by the observation data (see below). Second, for sufficiently small R_u, at the beginning of expansion, the term GM_u/R_u is sufficiently large for E to be neglected. (At $R_u \to 0, |E| \ll GM_u/R_u$.)

In the approximation adopted here, (2.4) has a simple solution,

$$(\tfrac{2}{3})R_u^{3/2} = (2GM_u)^{1/2}t_u + \mathrm{const}.$$

The time t_u will be reckoned from the moment when $R_u = 0$. It then follows that

$$\begin{aligned} R_u(t_u) &= (\tfrac{9}{2}GM_u)^{1/3}t_u^{2/3}, \\ R_u(t_u) &= (6\pi Gt_u^2)^{-1} \quad , \\ H(t_u) &= \frac{2}{(3t_u)} \quad . \end{aligned} \tag{2.6}$$

In theoretical works on cosmology, a more complex expression for Hubble's constant is found. This is connected with a limited validity of the Hubble law in the form (2.1), as has been used here. In our considerations, the non-relativistic approximation has been used. In the theory of general relativity, the curvature of space leads to a slight change in relation (2.1), resulting in a modification of the expression for Hubble's constant. However, we shall not need the exact expression; this simplified form of it will suffice and the age of the Universe can be calculated from the observed magnitude of Hubble's constant, $t_u \sim H^{-1}$.

2.4 The Universe: Open or Closed? (cf. Sect 4.12.1–2)

Let us touch further upon a vital question regarding the fate of our Universe: Is it open or closed? Unfortunately, no definite answer to this question can be given now. The average density of matter of the radiating regions of the galaxies, ϱ_g, has been measured with reasonable precision. It turns out that $\varrho_g \sim 10^{-30}\,\mathrm{g\,cm}^{-3} \sim 0.1\varrho_c$. The conclusion that our Universe is open might seem to follow from this; however, this conclusion is premature, and may even be wrong. The source of uncertainty is the lack of techniques for detecting "dark", nonradiating matter. Such matter could be associated with the following objects: stars of small mass, positioned at the periphery of the galaxies; neutron stars; black holes; neutrinos; gravitons; exotic particles etc. There are serious indications that the true density of matter is about one order of magnitude larger than that observed. This reasoning is based on the established fact of stationariness of the major part of galaxy clusters. To provide for this stationariness, the equilibrium condition has to be satisfied. From this condition, the mass of the galaxy clusters can be estimated. It is this mass which proves to be an order of magnitude larger than the figure determined from the radiation of the galaxy clusters. Although the "dark" objects are not identified, the above argumentation suggests that the true magnitude of the density of matter may be an order of magnitude larger than the apparent one and, hence, that[6]

$$\varrho \sim \varrho_c \quad . \tag{2.7}$$

So the question of the character of the evolution of the Universe cannot be definitely answered at present.

Our considerations were carried out in the framework of the Newtonian theory. The impression might arise that the theory of general relativity was a useful tool in constructing a model of the Universe, but redundant in interpreting the model. This is an erroneous opinion, however. Although the Newtonian theory yields results coinciding with the conclusions from the theory of relativity, the Newtonian approximation is not quite consistent. This inconsistency becomes evident if one tries to reconcile the basic cosmological postulates (homogeneity and isotropy) with the adopted model of a spherical Universe. It is obvious that there is no homogeneity and isotropy at the boundary of the sphere. (These properties are approximately fulfilled only in the regions inside the sphere.) In the theory of relativity, the space which includes the Universe is itself determined by the quantity ϱ; thus, the inconsistency is completely eliminated.

---

[6] This relation justifies the simplifying assumption $E = 0$ in (2.4).

This situation is best illustrated for the case of $\varrho > \varrho_c$ (closed Universe). The Universe is then a closed volume in a three-dimensional, generally non-Euclidean space. Though this volume is finite, there are no boundaries characteristic of a sphere in the above model. One can get some idea of a finite volume without boundaries by considering a spherical surface. Mathematically, a spherical surface is a two-dimensional space which is completely homogeneous and isotropic. (There is no preferred orientation on it.) And yet this space has no boundaries. Thus, a spherical surface is a classic example of a finite space without boundaries.

The above example gives, indeed, only *some* idea, of a finite volume without boundaries, the analogy being incomplete. As a matter of fact, a spherical surface is a two-dimensional entity, whereas the Universe occupies a three-dimensional volume. Unfortunately, it is not possible to give a pictorial example of a three-dimensional boundless entity.

It remains to be emphasized that the theory of general relativity provides a consistent and adequate description of the Universe, though some problems (cf. Sect. 2.8, 4.5.2) cannot be considered as finally solved.

2.5 A Hot Universe

There are two forms of existence of physical entities: matter and radiation. The analysis in the preceding few sections involved a tacit assumption that the Universe consists of matter only. Moreover, it was assumed that matter is cold, meaning that the kinetic energy is much smaller than the rest mass. This is the so-called model of a cold Universe which prevailed until the mid-sixties. But what followed then? What followed is one of the surprising stories the winding path of cosmology abounds in. In 1965 two young American radio astronomers, Penzias and Wilson, did a routine job of analyzing the noise in large antennae. Successively eliminating terrestrial and cosmic (Sun, Galaxy) noises and parasites,[7] they were not able to get rid of a weak background, with an equivalent temperature of $T \sim 3\,\text{K}$. This background, which could not be eliminated, could not be interpreted in a simple way. Its peculiarity lay in its isotropy. Without the knowledge of the two astronomers (who were later awarded a Nobel Prize), a new era in cosmology had thus begun.

The chance that brought radio astronomers and cosmologists together made it possible to give this radiation, now referred to as cosmic background radiation, the following interpretation. This background is an almost sym-

---

[7] One of the difficulties encountered by Penzias and Wilson was radio interference caused by pigeons roosting in their horn antenna.

bolic relic of the metagalactic radiation that was once predominant in the Universe. It is this interpretation which led to the now common belief that specifying the initial conditions for the expanding Universe as cold matter is wrong. In this connection Gamow's statement (1948) was also recalled, according to which the Universe must have been filled with radiation (the model of a hot Universe) at the beginning of expansion.

Hence, the model of a hot Universe leads − in the Newtonian approximation − to an analysis of the evolution of a sphere filled with radiation in its own gravitational field. Of course, this analysis is also based on (2.4) and (2.5) used in Sect. 2.3; only now the quantities entering them have to be given a somewhat different meaning. Let ε be the radiation energy density. From the equivalence of mass and energy it follows that

$$\varrho = \frac{\varepsilon}{c^2} \ . \tag{2.8}$$

Since the wave length λ is proportional to R and consequently the photon energy E_γ is inversely proportional to R, one has

$$\varepsilon = \frac{3N}{4\pi R_u^3} \frac{2\pi \hbar c}{\lambda} = a R_u^{-1} \ , \tag{2.9}$$

where N is the total number of photons and a is a constant. Using the relation $M_u = (\frac{4}{3})\pi \varrho R_u^{3}$ and (2.8), (2.9) equation (2.4) can be rewritten as

$$\frac{1}{2}\left(\frac{dR_u}{dt_u}\right)^2 - \frac{4}{3}\frac{\pi a G}{c^2 R_u^2} = 0. \tag{2.10}$$

(As before, we set $E = 0$.) The solution of this equation has the form

$$R_u = \left(\frac{32\pi}{3}\frac{aG}{c^2}\right)^{1/4} t_u^{1/2} \ . \tag{2.11}$$

The dependence of the "size" of the Universe on t_u is now weaker than in the model of a cold Universe, cf. (2.6). Making use of (2.9) as well as of the relation

$$\varepsilon = \sigma T^4 \ , \tag{2.12}$$

which is well known from the thermodynamics of black-body radiation (with $\sigma = \pi^2 k^4 / 15(\hbar c)^3$ being the Stefan-Boltzmann constant and k the Boltzmann constant), one gets the following relations:

$$\varepsilon(t_u) = \frac{3}{32\pi}\frac{c^2}{G t_u^2} \ ,$$

$$T_u(t_u) = \left(\frac{45}{32\pi^3 G}\frac{c^5 \hbar^3}{k^4}\right)^{1/4} t_u^{-1/2} \ , \tag{2.13}$$

and

$$H(t_u) = (2t_u)^{-1} \quad .$$

The last two relations are especially worthy of attention, for they can be checked by experiment.[8] Substituting the numerical values of the constants in the expression for the temperature T_u, one obtains a very simple relation,

$$T_u \sim \frac{10^{10}}{(t_u)^{1/2}} \tag{2.14}$$

(t_u being measured in sec and T_u in K.) Taking $t_u \sim 3 \times 10^{17}$ s, one finds that for the current epoch, $T_u \sim 10\,\text{K}$, which is in order-of-magnitude agreement with the measured value of $T_u \sim 3\,\text{K}$. This agreement can be considered very satisfactory, especially in view of the fact that this model of a hot Universe is rather idealized. Indeed, the evident existence of matter in the Universe has been neglected here. The real Universe has a mixed character consisting of both matter and radiation. In the current epoch, the energy density of matter, ε_m, prevails over the energy density of radiation, ε_r. Indeed, $\varepsilon_m \sim 10^{-9}\text{--}10^{-8}\,\text{erg cm}^{-3}$ while $\varepsilon_r \sim 10^{-12}\,\text{erg cm}^{-3}$; accordingly, $\varepsilon_r/\varepsilon_m$ amounts to $10^{-4}\text{--}10^{-3}$. Since the R_u-dependence of ε_m and ε_r in the model of a hot Universe is different, the ratio $\varepsilon_m/\varepsilon_r$ varies with R_u:

$$\frac{\varepsilon_m}{\varepsilon_r} \propto R_u \quad .$$

Consequently, there was a period in the history of the Universe when ε_m was about equal to ε_r. By using this equality one can easily recognize that this "turning point" epoch corresponds to $t_u^* \sim 10^3 - 10^4$ years $\sim 10^{11}$ s. At $t_u < t_u^*$, radiation was prevalent (the radiation-dominated era); at $t_u > t_u^*$, matter prevails (the matter-dominated era).

The value of $t_u \sim t_u^*$ plays an important role in the evolution of the Universe. It is during this period that the conditions necessary for the formation of galaxies are realized, cf. Sect. 8.3. Besides, the interaction between radiation and matter practically ceases at this time. The radiation only changes its wave length which, however, does not alter the spectrum shape that of course corresponds to the black-body radiation spectrum as given by the Planck distribution. Numerous thorough measurements of cosmic background radiation carried out in the long wave-length range of $\lambda \sim 1\,\text{mm}\text{--}10\,\text{cm}$ showed excellent agreement with the Planck spectrum for $T_u \sim 2.7\,\text{K}$.

---

[8] The first two equations are interrelated through (2.12) and thus are not independent.

This is a crucial substantiation of the validity of the basic assumptions underlying the model of a hot Universe.

The general ideas of an expanding Universe are also supported by the coincidence of the age of the Universe ($t_u \sim 10^{10}$ years) as determined from (2.13) with the measured age of many cosmic objects. In particular, the age of the Earth, which can be fairly precisely determined by measuring the concentration of various radioactive elements, is about 5×10^9 years. The age of the oldest stars has the same order of magnitude, about 10^{10} years. The occurrence of the cosmic blackbody radiation is proof that in the beginning of its expansion, the Universe was hot. The importance of this reasoning is so fundamental that it would be instructive to summarize some of the results of the investigations on cosmic background radiation.

As has been mentioned above, the energy spectrum of cosmic background radiation correlates well with Planck's distribution for the temperature $T_0 = 2.7\,\mathrm{K}$. It has been shown with high accuracy that cosmic background radiation is isotropic, which proves its extragalactic origin; were it of galactic origin, it would inevitably reflect the essential asphericity of the Galaxy and consequently, be anisotropic.

These facts are in quantitative agreement with the predictions of the hot Universe theory and confirm, in particular, one of the cosmological postulates, viz. that of the isotropy of the Universe. From the spectral distribution of the cosmic background radiation, its energy density ε_{r0}, the average photon energy E_{r0} and the photon concentration n_{r0} can be calculated. The resulting numerical values are: $\varepsilon_{r0} \sim 1\,\mathrm{eV\,cm^{-3}}$, $E_{r0} \sim 10^{-3}\,\mathrm{eV}$, and $n_{r0} \sim 10^3\,\mathrm{cm^{-3}}$.

In spite of their large concentration n_{r0}, the photons of cosmic background radiation have practically no effect on the processes occurring on the Earth. This is a consequence of the smallness of the photon energy E_{r0}.

2.6 Baryonic Asymmetry of the Universe

It is known with absolute certainty that the Earth as well as all objects on it consist of neutrons, protons, and electrons; there are absolutely no antiprotons on the Earth and in outer space. This phenomenon (which will be seen below to be of universal character) is known as *baryonic asymmetry*. Thorough measurements of the fluxes of solar cosmic rays emitted during solar flares have proven the complete absence of antiprotons in the interior of the Sun. An insignificant positron flux contained in solar cosmic rays is ascribed to the interaction of the primary solar cosmic rays (protons, electrons, and photons) with matter leading to the conclusion that positrons are practically absent on and within the Sun.

A similar approach can be used to prove the absence of antimatter in the entire Galaxy. At present, it can be considered a firmly established fact that the majority of cosmic rays with energy above 1 GeV originate within the Galaxy. Detailed investigations of the composition of galactic cosmic rays have also shown that insignificant fluxes of positrons and antiprotons therein are readily explained by secondary processes. Hence, there is no antimatter in the Galaxy either. It is more difficult to give a direct proof of the absence of antimatter in other galaxies or in the Universe as a whole. The difficulty is that at present, the only channel of information on the processes occurring in other galaxies is electromagnetic radiation. But antimatter emits photons in absolutely the same way as matter does. Thus the radiation spectra of other galaxies do not indicate whether they consist of matter or antimatter.

Proving universal, global charge asymmetry is of fundamental importance, for this nontrivial fact stimulates questioning its origin. Indeed, supposing that charge symmetry holds, it would express a profound relationship between the microcosm and the macrocosm. After all, the Universe consists of elementary particles, and it would be natural to expect a coincidence of (or at least an agreement between) the laws ruling the microcosm and those governing the Universe. As far as the microcosm is concerned, until the late 1950s or early 1960s[7] a belief prevailed in the theory of elementary particles that particles and antiparticles are identical, except for the charge sign. This can be visualized in the following thought experiment. Imagine a closed system consisting of particles with both positive and negative charge; inversion of the sign of the particle charges will then not alter the motion of a test particle whose charge is also inversed.

The principle of charge symmetry was given a more profound justification in the Dirac equation, which is the basis of quantum electrodynamics. In the framework of this theory, particles and antiparticles are identical (except for the charge sign). The principle of charge symmetry on the microscale thus appeared unshakeable and it seemed quite appropriate to extend it to the whole Universe. This was a mainstream tendency in cosmology until the mid-sixties. The hypothesis that the galaxies consisting of matter and those consisting of antimatter ("antigalaxies") occur in the Universe in equal numbers was quite popular at that time. In its simplest version, excluding all structures but galaxies and antigalaxies, the hypothesis was short-lived, however.

Galaxies emit – besides electromagnetic radiation – particles as well. Our Galaxy emits protons and electrons; an antigalaxy would have to emit antiprotons and positrons. Were, for instance, the nearest neighbour galaxy, the Andromeda Nebula, an antigalaxy, it would emit an appreciable number

---

[7] What happened then is described below.

of positrons; in intergalactic space, these positrons would meet the electrons emitted by our Galaxy. Annihilation would result according to the reaction $e^+ + e^- \rightarrow 2\gamma$ (γ denoting a gamma-quantum).

For the most part, this reaction occurs at small kinetic energies of the electrons and the positrons. Therefore, the energy E_γ of each of the quanta created in this annihilation reaction should be given by

$$E_\gamma \sim m_e c^2 \sim 0.51\,\text{MeV} \quad .$$

Such quanta penetrate the Galaxy almost without let or hindrance and can reach the outskirts of the Earth's atmosphere. In the late 1960s γ-ray astronomy, which permits recording γ-quanta with energies down to 0.1 MeV, began to flourish. An intensive excess radiation with the energy $E_\gamma \sim 0.5$ MeV would certainly have been noticed even in the first, rather rough measurements. No noticeable intensity excess over the background has been detected in the region of $E \sim 0.5$ MeV, however, which is in contradiction to the simplest hypothesis of the existence of antigalaxies.

To save this hypothesis, an assumption was put forward that additional structural elements exist, viz. magnetic "walls", separating galaxies from antigalaxies. However, this comparatively complex construction was not able to rescue the hypothesis of the existence of antigalaxies either. First, it turned out that no magnetic walls can keep electrons and positrons completely separate. Leakage of particles through the hypothetical magnetic walls would lead to excess radiation with $E \sim 0.5$ MeV, which could be detected by the more sophisticated devices in use in γ-ray astronomy. Second, the energy of the magnetic walls would be so large that their existence would distort the observed isotropy of the cosmic background radiation. Finally, in the magnetic wall hypothesis, the question of the cause of charge asymmetry in the Universe was just replaced by another: Which processes lead to magnetic wall formation? Thus in the middle of the 1960s, a general mood of frustration prevailed with regard to charge symmetry. It appeared that charge asymmetry had arisen at the instant of the birth of the Universe, i.e., at the moment in which it went through a singular state[8] Now cosmology took one of its miraculous turns, repeatedly mentioned above.

In 1967, a concept explaining the charge asymmetry of the Universe was put forward by A.D. Sakharov. At first, the idea was mistrusted, for it contained cardinally new elements. However, by the end of the 1970s

---

[8] It should be mentioned that the cosmological singularity, i.e., the moment when the energy density and the temperature become infinite, often plays the role of a carpet under which "cosmological rubbish" – phenomena not explainable within the standard model of the Universe – is swept.

and in the early 1980s, it had gained general acceptance in connection with the development of interaction unification (cf. Sect. 1.5) and is now quite popular. This idea is based on two hypotheses:

i)   the baryon number is not a strictly conserved quantity and
ii)   the charge symmetry is not an absolutely exact symmetry.

Let us explain some details of these hypotheses. The baryon number gives the difference between the number of baryons and that of antibaryons. In the simplest and most important case, namely, under usual terrestrial conditions, the baryon number is equal to the total number of protons and neutrons. Nonconservation of the baryon number indicates, in this case, the possibility of proton decay. As a matter of fact, the proton decay hypothesis (cf. Sect. 1.6) was proposed in cosmology long before the advent of the theory unifying three interactions!

Let us now turn to the question of charge symmetry. It was mentioned above that particles and antiparticles are symmetrical with respect to interactions. However, this statement has been proven for the strong and the electromagnetic interactions only. As early as 1956, it was demonstrated that weakly interacting neutrinos and antineutrinos differ not in their charge (both being electrically neutral, of course), but in a new quantum number, *chirality.* Chirality is the relative orientation of the vectors of spin and momentum. For an antineutrino, the directions of spin and momentum are coincident while for a neutrino, the two vectors have opposite directions. Still, the basic rule of particle transformations based on the charge symmetry principle seemed to be inviolable: a particle a decays into particles b and c with the probability $W_{a \to b+c}$, identical with the probability $W_{\bar{a} \to \bar{b}+\bar{c}}$ of decay of the antiparticle $\bar{a}$ into the antiparticles $\bar{b}$ and $\bar{c}$. Fitch and Cronin (1964) cast doubt upon this statement, showing in fact that this rule is violated in the particular case (the only one as yet) of K-meson decay. Although the difference between the probabilities $W_{a \to b+c}$ and $W_{\bar{a} \to \bar{b}+\bar{c}}$ was very small (about 0.1 %), the very fact of charge symmetry violation was of paramount importance, and Fitch and Cronin were awarded a Nobel Prize.[9]

Now it is time to come back to our subject, baryonic asymmetry. It is clear that both conditions together, viz. nonconservation of the baryon number and charge symmetry violation, are necessary for the baryonic asymmetry of the Universe to occur. Nonconservation of the baryon number alone, at $W_{a \to b+c} = W_{\bar{a} \to \bar{b}+\bar{c}}$, would only imply that protons and antiprotons decay in the same way, the charge symmetry of the Universe not being

---

[9]   It is worthwhile mentioning that the charge symmetry violation was so weak that just after the lecture by Fitch and Cronin, a hypothetial "explanation" of the effect caused by a fly, accidentally trapped in the device, was speculated upon

violated. Conversely, charge symmetry violation, with conservation of the baryon number, would not affect the proton-antiproton balance.

Despite the apparent simplicity of the explanation just given, the baryon number nonconservation hypothesis did not seem so obvious in the late 1960s. This interpretation of baryonic asymmetry appeared in a completely new light in connection with the theory of grand unification (cf. Sects. 1.5–6), in which nonconservation of the baryonic charge is quite natural.

The grand unification theory also offered the possibility of accounting for the charge symmetry violation for elementary particles. In particular, the approach proposed by Kobayashi and Maskawa was to become quite popular; it explained charge symmetry violation for a minimum number of fundamental particles.[10] The minimum number of leptons (or quarks) in this approach is equal to six, in accord with modern evidence. Five leptons $(e, \mu, \tau, \nu_e,$ and $\nu_\mu)$ as well as five quarks have been revealed to date, but nobody doubts the existence of a sixth lepton (τ-neutrino) and an associated quark. It is interesting to note that only three quarks were known at the time Kobayashi and Maskawa published their approach. Heavy particles (possessing charm and flavor) were detected later on.

Let us consider some particular aspects of the interpreation of the charge asymmetry of the Universe in terms of the grand unification theory. It predicts the existence of heavy X-bosons with mass $\sim 10^{15}$ GeV and charge $\pm\frac{2}{3}e$ or $\pm\frac{4}{3}e$ (cf. Sect. 1.5.2). The X-bosons can decay, e.g., according to the reactions

$$
\begin{aligned}
X^+ &\to \bar{q}_1 + e^+ \quad , \\
X^- &\to q_1 + e^- \quad , \quad \text{and} \\
X^+ &\to 2q_2 \quad .
\end{aligned}
\tag{2.15}
$$

The charges of the quarks q_1 and $\bar{q}_1$ equal $\pm\frac{1}{3}e$; that of q_2 equals $\frac{2}{3}e$. In the next step, two quarks q_2 combine with a quark q_1 to form a proton. It is clear that in all of the reactions (2.15), the baryon number cannot be conserved: particles with various baryon numbers are generated as a result. In the first two reactions, the baryon number equals $\pm\frac{1}{3}$ while in the last one, it equals $\frac{2}{3}$. Furthermore, it follows that because of the charge symmetry violation, the probabilities of the occurrence of the first two decays are different. As a consequence, the number of the quarks q_1 created exceeds the number of the antiquarks $\bar{q}_1$, giving rise to asymmetry. This process is especially effective at very early stages of the expansion of the Universe $t_u \sim 10^{-40}$–10^{-30} s).

This approach has been subject to numerous theoretical checks. A quantitative measure of baryonic asymmetry is the ratio $S = n_\gamma/n_p$ where

---

[10] This approach was recognized in the Nobel lectures by Fitch and Cronin.

n_γ and n_p denote the photon and the proton abundance, respectively. During a major part of the evolution of the Universe, this ratio remains constant. The quantity S characterizes the ratio of the concentration of the neutral components of matter to that of the charged components. According to observational data, S amounts to 10^8–10^{10}. Theoretical estimates made on the basis of the grand unification theory yield with certainly a lower limit for S, $S > 10^5$; most plausible are the values of $S \sim 10^9$–10^{10}. These estimates are an indirect (and, up to date, the only) confirmation of the grand unification theory.

It should be noted in conclusion that the elegant picture of the origin of the baryonic asymmetry of the Universe outlined above would be more complete if proton decay were revealed (cf. Sect. 1.6).

2.7 Cosmologic Nucleosynthesis of Helium

The efforts of many physicists and astrophysicists during the two decades following the pioneering work of Gamow, Alpher, and Herman (1948) have led to a self-consistent and generally accepted theory of the synthesis of chemical elements. The paths that led to this theory were far from straight and abounded in guesses and fallacies. Its success was promoted by remarkable progress in accumulating observational data; they can be summarized as follows:

1. The helium distribution in the Galaxy and in the Universe is roughly uniform.
2. The relative helium concentration in the Universe is about 6 %, corresponding to the abundance of helium by mass of 25 % (the helium nucleus, an α-particle, being 4 times heavier than a proton).
3. Heavier elements are distributed in the Universe nonuniformly, with higher concentrations in the proximity of stars and star clusters.

Besides these experimental facts, two simple but very significant considerations influenced the development of the theory of nucleosynthesis. The first is that such a prominent abundance of helium could not stem from nuclear synthesis occurring within stars (cf. Sect. 2.9.2). To account for such a high percentage, an excessively large energy of fusion would have to be assumed for interstellar helium nucleosynthesis. The estimated value of the energy released and the observational results differ by more than one order of magnitude.

The other consideration is connected with a peculiar feature of the periodic system which does not contain a stable element with the atomic number $A = 5$. This rules out the simplest imaginable variant of synthesis of the elements heavier than helium by fusion of two particles (e.g., fusion of an α-particle and a neutron). The fusion of two helium nuclei to produce ^{7}Li according to the reaction ^{4}He$+^4$He $\rightarrow ^7$Li$+p$ is relatively inefficient, due to a large value of the Coulomb barrier of the participating helium nuclei. Estimates showed that fusion reactions of complex elements do not play a significant role in the processes of nucleosynthesis at early stages of the expansion of the Universe. More complex variants (e.g., three-particle processes in the formation of heavy elements at an early stage of cosmological expansion) are also ineffective, because of their slow rate. Thus, a fundamental conclusion follows from the experimental and theoretical data: α-particles arise primarily at early stages of cosmological expansion, while heavier elements are formed in the process of stellar evolution. (In this section, the former process is considered; the theory of synthesis of heavy elements in stars will be treated in Sect. 2.9.2.)

A main chain in cosmic nucleosynthesis is the deuteron production in a reaction

$$p + n \rightarrow d + \gamma \ . \tag{2.16}$$

Following this reaction, the deuterons convert to α-particles via various channels, e.g. via the reactions

$$d + d \sim t + p, \quad \text{and} \quad t + d \rightarrow \alpha + n \ . \tag{2.17}$$

Here, t denotes a tritium nucleus (made up of one proton and two neutrons).

The kinetics of the helium production processes are basically determined by two factors: (i) the concentration of neutrons required for reaction (2.16) to proceed effectively and (ii) the cosmic background radiation inhibiting reactions (2.17), due to deuteron photodissociation, $\gamma + d \rightarrow p + n$. Hence, to evaluate the resulting α-particle concentration, the variation of the neutron concentration, and the effectiveness of photodissociation during reactions (2.16) and (2.17) should be given detailed consideration. Here we confine ourselves to semi-quantitative estimates.[11]

First of all we consider qualitatively the variation of the relative neutron abundance with temperature T_u of the Universe or with time t_u, reckoned from the beginning of the expansion. The latter two quantities are interrelated through (2.14) so that, e.g., by fixing T_u, the time t_u is also specified.

For very high temperatures, $T_u \gtrsim m_p c^2 / k \sim 10^{13}$K, the difference of the properties of protons and neutrons does not come to bear. Consequently,

---

[11] A detailed account of the theory of helium nucleosynthesis is given in : P.J.E. Peebles: *Physical Cosmology* (Princeton Univ. Press, Princeton, N.J. 1971)

the ratio of their abundancies, n_n/n_p, is about equal to unity. However, as the temperature T_u approaches the value $\Delta m_N c^2/k$ (where $\Delta m_N = m_n - m_p$), the excess mass of the neutrons as compared to the proton mass, $\Delta m_N \sim 1.3\,\text{MeV}$, comes into play. A difference in the abundances will be caused by the fact that the neutrons and protons are in thermodynamic equilibrium, the concentration being determined by the Boltzmann factor. The relative concentration (abundance) is then $n_n/n_p \sim \exp\left(-\Delta m_N c^2/kT_u\right)$. While this factor is close to unity for $kT_u \gg \Delta m_N c^2$, it starts drastically decreasing with decreasing temperature for $kT_u \lesssim \Delta m_N c^2$. This line of reasoning depends on the existence of statistical equilibrium which implies the occurrence of reactions with mutual conversions of protons and neutrons.

A detailed analysis shows that in the range $kT_u \lesssim \Delta m_N c^2$, the processes determining such mutual conversions are represented by the following reactions caused by the weak interaction:

$$\bar{\nu} + p \rightleftarrows e^+ + n, \quad \nu + n \rightleftarrows e^- + p \ . \tag{2.18}$$

These reactions determine the neutron abundance in the time interval $0.1\text{s} < t_u < 1\,\text{s}(10^{10}\,\text{K} < T_u < 3 \times 10^{10}\,\text{K})$. At $t_u \gtrsim 10\,\text{s}$, the decreased nucleon concentration makes it possible for neutrinos to cross the Universe with practically no interaction. The balance is then disturbed, and the neutron-proton abundance, n_n/n_p, remains in this time almost constant until the nucleosynthesis era sets in, i.e., until the reaction of deuteron dissociation by the photons of the cosmic background radiation is rendered ineffective. Then the period of nucleosynthesis begins $(t_u \sim 200\,\text{s}, T_u \sim 10^9\,\text{K})$: all neutrons convert to α-particles via reactions of the type (2.16) and (2.17). In Fig. 2.4, the solid line

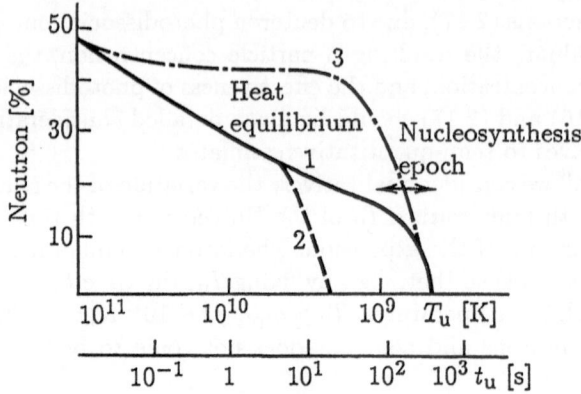

Fig. 2.4. The dependence of the relative neutron abundance, $n_n/(n_n + n_p)$, on time t_u calculated for (1) a realistic value of α_w; (2) for $\alpha'_w = 10\alpha_w$; (3) for $\alpha'_w = 0.1\alpha_w$

represents the ratio $n_n/(n_n + n_p)$. It follows from the plot shown there that the relative abundance of helium by mass is about equal to 25 %, which is in good agreement with the observational data. This accord, together with the absence of any alternative explanation for the large abundance of helium, signifies a reasonable success of the hot Universe model, valid for the times up to $t_u \sim 0.1$–1 s.

To conclude, one interesting and very important fact should be noted. The ratio of the abundances of α-particles and protons has the order of unity; this value results from a bizarre interplay of various factors that may shift it in both directions. In our discussion, instability of a free neutron has not been taken into account. As is well known, a free neutron can decay via $n \rightarrow p + e^- + \bar{\nu}$, its lifetime being $\tau_n \sim 10^3$ s. This time exceeds by far the time of efficient nucleosynthesis so that neutron decay can be neglected for practical purposes. The consideration of this factor would introduce a correction in the ratio n_α/n_p amounting to about 10 %. The situation would be radically different if the parameters determining τ_n had somewhat different values. It is known from the theory of β-decay that

$$\tau_n \propto \frac{1}{\alpha_w^2 (\Delta m_N)^5} \quad .$$

If, for example, a hypothetical value α_w' of the weak interaction coupling constant exceeded the value α_w for the Universe by approximately one order of magnitude, the neutron lifetime would have the value $\tau_n \sim 10$ s, and at the time at which nucleosynthesis is to occur, there would be practically no neutrons available. In this case, the abundance of helium would be zero. A similar situation would arise if Δm_N were increased by the factor of about two.

In the opposite case, i.e., for a value of α_w reduced by one order of magnitude, reactions (2.18) would be rendered ineffective. Statistical equilibrium caused by the weak interaction would not be achieved, and the proton and neutron abundances would be approximately equal up to the nucleosynthesis era (the dot-and-dash line in Fig. 2.4). In such a case, practically all protons and neutrons would convert to α-particles via the primary reaction (2.16). A detailed analysis of these hypothetical situations is given below, in Sect. 3.3.

2.8 The Origin of Galaxies

In his book entitled *The First Three Minutes**, S. Weinberg writes: "The theory of the formation of galaxies is one of the great outstanding problems of astrophysics, a problem that today seems far from solution." This viewpoint, expressed in 1977 by the celebrated American scientist, is still shared by many physicists and astrophysicists. Therfore we shall not give an account of the very complex models of galaxy formation which differ greatly in their approach. Instead, we shall concentrate on the analysis of the basic difficulties in constructing a theory of galaxy formation as well as on the derivation of certain conditions which are necessary but − alas − not sufficient for galaxy formation.

Principal Structural Elements of the Universe

The characteristics of the principal structural elements of the Universe are summarized in Table 2.1. It should be noted that the date listed in Table 2.1 do not completely represent the whole variety of structural elements. An important feature not reflected in Table 2.1 is the smoothness of transition from stars to galaxies and from galaxies to galaxy clusters. Besides the giant galaxy clusters whose characteristics are given in Table 2.1, there are clusters in the Universe which consist only of a few galaxies. Within galaxies, star clusters exist. The diversity of the structural elements of the Universe is thus represented not only by the strictly demarcated categories "stars-galaxies-galaxy clusters" but also by a continuous transition from single stars to star agglomerations and further to galaxies and galaxy clusters. However, the values of the size and mass of the structural elements given in Table 2.1 are those predominant among the multitude.

Table 2.1. Characteristics of the principal structural elements of the Universe

Structural elements	Size		Mass	
	$[pc]^*$	[cm]	$[M\odot]$	[g]
Galaxy clusters	10^6–10^7	10^{24}–10^{25}	10^{15}	10^{48}
Galaxies	10^3–10^5	10^{21}–10^{23}	10^{10}–10^{12}	10^{43}–10^{45}
Stars	10^{-8}–10^{-6}	10^{10}–10^{12}	10^{-1}–10^2	10^{32}–10^{35}
The Sun	5×10^{-8}	1.4×10^{11}	1	2×10^{33}

$1\,pc = 3 \times 10^{18}$ cm

* S. Weinberg: *The First Three Minutes* (Basic Books, N.Y., 1977).

Physical Principles of Galaxy Formation

Let us now consider the physical principles underlying the theory of galaxy formation. The basis of all models is the gravitational instability of which already Newton was aware. This instability is associated with the attractive character of the gravitational forces. Hence, should fluctuation occur in the initially homogeneously distributed matter, causing a density excess over the average density, such a fluctuation will generally result in a contraction of the matter. This idea was quantitatively worked out in the 1920s by the British astrophysicist, J. Jeans. The analysis was based on the general laws of mechanics which have been used in Sects. 2.1 and 2.5 in modelling the evolution of the Universe. If $E_k > E_p$ (E_k and E_p denoting the kinetic and the potential energy of an element of matter, respectively), then no steady-state is possible: the matter will undergo recession. In the case under consideration, this corresponds to the damping of the fluctuation. Under the condition $E_k < E_p$, contraction will occur. At approximate equality, $E_k \sim E_p$ will lead to an equilibrium configuration. The potential energy of a pair of particles (p, e^-) is obviously

$$E_p \sim \frac{GMm_p}{R},$$

where M and R are the characteristic values of the mass and the radius of the density fluctuation. The situation regarding the kinetic energy is somewhat more complex. Kinetic energy takes two forms: thermal and radiative. The order of magnitude of the thermal energy of a particle pair is kT; the radiative energy of the pair (p, e^-) equals SE_γ, where $S = n_\gamma/n_p$ and E_γ is the average energy of a quantum of radiation. The equilibrium condition can be then expressed as

$$\frac{GMm_p}{R} \sim kT \tag{2.19}$$

(radiation pressure being neglected) or

$$\frac{GMm_p}{R} \sim SE_\gamma \tag{2.20}$$

(pressure being determined by radiation).

Contraction occurs if the left-hand side of (2.19) or (2.20) exceeds the right-hand side. It might seem that these conditions could be easily satisfied, provided that arbitrary values of M and R are appropriately chosen; this is exactly what was believed in Jeans' times. However, the discovery of the expansion of the Universe and cosmic background radiation led to complexities in finding the equilibrium conditions.

Let us first consider the influence of the expansion of the Universe. As follows from Hubble's relation, the particles within the fluctuation are moving with respect to each other; expansion has the effect of pulling them apart. The expansion of the Universe is thus a factor impeding the contraction of the fluctuation. From (2.4) and (2.5), it is easily seen that the relative velocity of the particles caused by expansion increases with decreasing average density ϱ. This dependence is a consequence of a decrease of the net attraction force with decreasing ϱ, which prevents matter from recessing.

Jeans assumed that the primary source of gravitational perturbations are natural, statistical density fluctuations inherent in any homogeneous matter. However, it was shown by E.M. Lifshitz in 1946 that such fluctuations will inevitably decay, due to the expansion of the Universe. Consequently, the formation of galaxies cannot be accounted for by a model of a homogeneous Universe subjected to statistical fluctuations only. Some embryo fluctuations have to be introduced "by hand" which are then to be matched to the numerous observational data available on galaxies. One encounters here a tendency that contradicts, in a sense, the spirit of the Friedmann model, based on a clear-cut formulation of the initial conditions, viz. those of isotropy and homogeneity. The principles of isotropy and homogeneity are unique in their simplicity whereas disturbances or deviations from these postulates are diversiform. This is one of the reasons for the variety of galaxy formation models and for their incompleteness.

The situation is aggravated by the cosmic background radiation. Its direct influence is obvious: it prevents matter from condensating (cf. (2.20)). The temperature of the Universe, T_u, decreases, however, and as a matter of fact, one can expect that at a certain moment, the effect of radiation on condensation will become comparatively small. A sufficient condition for cosmic background radiation not to prevent galaxies from being formed can be readily written down. Indeed, particle velocity being smaller than the velocity of light , the condition

$$\frac{GM}{R} < c^2$$

is obviously fulfilled. Combining it with the Jeans relation (2.20) yields the desired condition for galaxy formation to be feasible:

$$SE_\gamma < m_p c^2 \quad . \tag{2.21}$$

For the present epoch this condition is fulfilled safely (with a margin of 4 to 5 orders of magnitude).

Cosmic background radiation has an indirect, very "dangerous" influence. As already mentioned, its high degree of isotropy is an excellent substantiation of the Friedmann model. But it is this isotropy which poses a serious limitation on the choice of initial disturbances. It can be shown that too large disturbances would lead to the violation of the isotropy of the cosmic background radiation, while too small ones, for which the Jeans conditions do not hold, would decay. This dilemma is one of the major difficulties in constructing a theory of galaxy formation.

Theories of Galaxy Formation

At present, there are several scenarios, to account for the formation of galaxies. In one of them it is assumed that the initial disturbances are of turbulent character and arise at the very beginning of the expansion of the Universe; i.e., they are intrinsically connected with the singular state of the Universe. The difficulty of this model is that eddy motion is damped in the course of the expansion. Therefore, disturbances of this kind will be effective enough only provided that large amplitudes of initial disturbances are admitted, which would, as just mentioned in the last section, impair the isotropy of the cosmic background radiation.

Another scenario, the *adiabatic theory of the formation of galaxies,* is based on the assumption that primordial disturbances are caused by fluctuations of the total density ϱ, the ratio $S = n_\gamma/n_p$ remaining invariable, i.e., the fluctuations of matter and radiation being phase-coincident. Finally, in the so-called *entropy theory*, perturbations are associated with fluctuations in the magnitude of S.

This seemingly insignificant difference in the character of disturbances leads to totally different pictures of the evolution of large-scale elements of the Universe. In the *adiabatic theory*, large structural elements with mass about equal to that of galaxy clusters arise first. Later on, in the process of the evolution of the Universe, their fragmentation leads to the formation of galaxies with mass of the order $10^5\,M_\odot$, followed by further decay into stars. The *entropy hypothesis* predicts that formations with mass $\sim 10^6\,M_\odot$ arise first. Later on they condense into galaxies or galaxy clusters. This process is also accompanied by the decay of the primary formations into stars.

For this reason, we shall refrain from further discussion of the theory of galaxy formation and dwell only on the derivation of certain conditions necessary for galaxy formation. These conditions will come in handy below (cf. Sect. 3.3). To make the derivation clear, we shall consider the thermal radiation accompanying the compression of matter in a fluctuation. This radiation stems from the Coulomb interaction between particles (bremsstrahlung). It is easily recognized that bremsstrahlung is especially pronounced if the matter is in the ionized state, and the bremsstrahlung is associated with the free electron-proton interaction. If the matter is ionized, the radiation "scatters" the arising density fluctuations. The intensity of the bremsstrahlung is substantially reduced if the matter is in the bound atomic state, i.e., if it is formed after the recombination of the plasma. There is a simple explanation of this effect: in the bound state, the Coulomb fields of protons and electrons are largely compensated for (screening). An estimation of the mass of fluctuations gives unreasonably large values. This implies that there must be a period in the evolution of the Universe when electrons and protons combine to form hydrogen atoms. Recombination occurs when the temperature satisfies the condition

$$kT \lesssim 0.1\varepsilon_H \sim 1\,\mathrm{eV} \tag{2.22}$$

where ε_H is the electron binding energy in a hydrogen atom. Condition (2.22) approximately corresponds to the temperature $T_u \sim 10^4$ K. Using relation (2.14) one can easily estimate that recombination in the Universe occurs at the time $t_u \sim 10^{12}$ s. It is interesting to note that at about the same time condition (2.21) for matter to prevail over radiation is first fulfilled.

Hence, the time $t_u \sim 10^{12}$ s is physically distinguished in the history of the Universe, for it is the time at which the conditions for the formation of large-scale structures begin to hold. One might think that it is this period when galaxies must be nucleated. However, observational data contradict this simplistic assumption. Currently it is believed that galaxies are formed appreciably later, at $t_u \sim 10^{16}$ s, during a period which cannot be regarded as physically preferred with respect to any particular conditions.

There is some hope that the discrepancy arising can be eliminated by the *hierarchical scheme of galaxy formation*. According to this, large-scale structures arise first, decaying into galaxies within 10^{16} s. In other schemes, it takes small-scale structures a time $t_u \sim 10^{16}$ s to transform to large-scale entities, the galaxies.

This brief excursion into the theory of galaxy formation could be concluded by Weinberg's sentence cited at the beginning of Sect. 2.8: There is no established and commonly accepted theory of galaxy formation. This somewhat pessimistic statement reflects the contradiction between the basic cosmological postulates (homogeneity, isotropy) and the existence of large-scale structural elements which violate the homogeneity of matter in the Universe. Perhaps this difficulty indicates that the Friedmann model is only an approximation needing modification. A shortcoming of Friedmann's model is also recognized in connection with the familiar problem of the horizon of the Universe. The essence of the problem is that only those points can be connected via light signals whose spacing is small compared to the size of the Universe. There is no causal relationship between two points in the Uni-

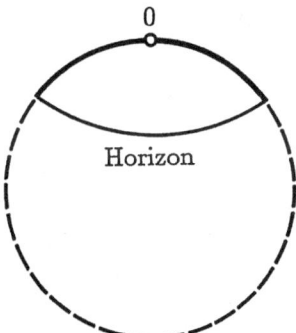

Fig. 2.5. The existence of a horizon in the Universe. A signal emitted from point 0 can only reach the region bounded by the solid line

verse remote from one another (Fig. 2.5). This is in obvious contradiction to a global characteristic of the Universe, viz. the amazing isotropy of cosmic background radiation: How could one part of this radiation become aware of the state of the radiation in the region not causally connected with it? Thus the problem of the horizon is a serious challenge to the Friedmann model of the Universe. This problem as well as the approaches to its solution are analyzed in Sect. 4.5.2.2.

2.9 Stars

Stars are the main structural constituents of the Universe. In this section, stationary (or, more strictly, quasi-stationary) stars are considered. By definition, the mass and the radius of the stationary stars do not significantly vary within a time comparable with the age of the Universe, t_u. The overwhelming majority of the stellar population of the Galaxy (above 99 %) are stationary stars.

2.9.1 Classification of Stars

Observing the stars was one of the major concerns of astronomy until the beginning of the 20th century. Very rich material has been accumulated as a result of the observations carried out over many centuries. This material has been subjected to thorough classification and systematization.

Hertzsprung-Russel diagrams offer the most condensed presentation of this vast body of information. The abscissa of such a diagram gives the color of a star, or more precisely, its spectral type. As is well known, a color in the optical range corresponds to the wave length of radiation or to a characteristic photon energy E_γ. Blue color corresponds to the smallest, and red color to the largest, wave length. Alternatively, blue corresponds to the largest value of E_γ or of T_s, the characteristic temperature of the radiating region. For the red, the quantities E_γ and T_s have the smallest values.

In the Hertzsprung-Russell diagrams the wave length usually increases along the abscissa; those near the origin correspond to relatively high temperature. Increasing the wave length of stellar radiation, i.e., moving to the right along the abscissa, from blue to yellow and further to red stars, corresponds to decreasing T_s.

The ordinate axis represents the luminosity of the star, i.e., the energy released per unit time (in $\mathrm{erg\,s^{-1}}$). In astronomy, a specific system of units is used; the Hertzsprung-Russel diagram in Fig. 2.6 (described below) is plotted in these units. The luminosity of stars is given in the logarithmic scale (as the absolute stellar brightness M). Usually a color is characterized by a spectral class with which a definite characteristic temperature T_s is associated. This classification is presented in Table 2.2.

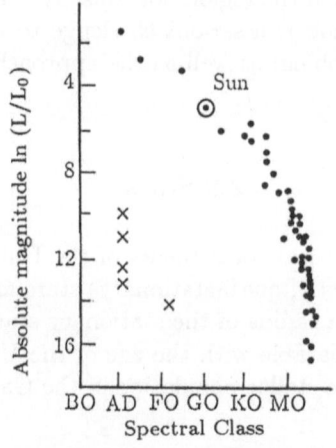

Fig. 2.6. The Hertzsprung-Russel diagram for the stars from within 5 pc from the solar system

Fig. 2.6 represents a typical Hertzsprung-Russell diagram for a star within a distance of 5 pc from the Earth. There are two distinct regions in the diagram in which points corresponding to the stars of various classes are concentrated. One is a band that spreads from the upper left corner to the lower right corner of the diagram. It is associated with the stars of the main sequence, i.e., with the overwhelming majority of all observable stars. The upper left part of the band corresponding to the stars of

Table 2.2. Classification of spectral class by temperature

Spectral class	Temperature [K]
B0	25 000
A0	11 000
F0	7 600
G0	6 000
K0	5 100
M0	3 600

the main sequence represents large blue stars and the lower right part, the red dwarfs. They are physically distinguished from the rest of the main sequence stars by the practically complete absence of thermonuclear reactions (see below). The Sun is positioned roughly in the middle of the band of the main sequence stars; it is an average (yellow) star, unremarkable from the astrophysical viewpoint. The points corresponding to the white dwarfs lie in the left part of the diagram.

The neutron stars are not represented in this diagram. They emit radiation intensively only within a relatively short period of time, in the order of 10^4–10^5 years. Accordingly, the vast majority of neutron stars are dark objects. In the vicinity of the solar system, there are no radiating neutron stars. It should also be noted that young neutron stars could hardly be represented in a Hertzsprung-Russell diagram which includes remote stars as well. The reason is that young neutron stars emit radiation in a rather broad wave lenth range (from the radio to the gamma-ray frequency) and therefore this radiation cannot be represented by a single color or temperature.

Table 2.3 is a compilation of the data for typical stars of the various classes. As a typical thermonuclear star of the main sequence, the Sun has been chosen.

Table 2.3. Typical stars of various classes

Stellar class	Mass [g]	Radius [cm]	Luminosity [erg s^{-1}]	Surface temperature [K]
The Sun (a typical star)	2×10^{33}	7×10^{10}	4×10^{33}	5×10^3
Red dwarfs	10^{32}	10^{10}	10^{30}	2×10^3
White dwarfs	10^{33}	5×10^8	10^{30}	10^4
Neutron stars	10^{34}	10^6	–	–

2.9.2 Biography of a Star

We shall now recount the life story of a typical radiating star. As will be seen, the evolution of such a star is closely related to the existence of white dwarfs and neutron stars.

The Energy Problem

Let us first dwell on the major problem of stellar evolution, viz. that of reconciling a large energy release (luminosity) with the stationary nature of stars. A high luminosity is necessary for maintaining stationarity. Indeed, the basic physical factor providing for stellar stability is the equality of the potential energy associated with the gravitational attraction and the kinetic energy associated with the thermal motion that produces a pressure opposing gravitational contraction. The corresponding balance equation was given earlier (cf. (2.19) in Sect. 2.8.2). For equilibrium, high pressure and temperature, and consequently, luminosity, are required.

Elementary estimates demonstrate, however, that the most natural energy source, gravitation, does not solve the energy problem. In fact, the total gravitational energy of a star amounts to

$$E_g \sim \frac{GM_s^2}{R_s} \sim 10^{48}\,\text{erg}$$

(cf. the parameters of the Sun given in Table 2.1). The gravitational lifetime of the star is then

$$t_s \sim \frac{E_g}{L_s} \sim 10^{14}\,\text{s} \quad,$$

(L_s denoting the luminosity of the star) which is smaller than the true lifetime of a star by a factor of 1000. This problem had already been recognized at the beginning of the 20th century, but its solution only took shape in the post-war years, thanks to the efforts of many theorists. The most significant contribution to solving the problem was made by H. Bethe. The solution was found not as a result of an ingenious guess, but rather as a logical consequence of the development of nuclear physics. The basic statement is formulated as follows: The principal source of the luminosity of the Sun as well as of the typical stars of the main sequence is the energy of thermonuclear fusion. By nuclear transformation of light elements into heavier ones, the energy $\sim (10^{-3} - 10^{-2})M_s c^2$ is released; this energy is quite sufficient for maintaining stellar luminosity for a long time. For example, the thermonuclear energy stored in the Sun amounts to approximately 10^{52} erg, which is enough to provide for its luminosity for 10^{11} years.[12]

Let us consider the hydrogen-helium cycle of fusion reactions:[13]

$$p + p \rightarrow d + e^+ + \nu + 0.5\,\text{MeV} \quad ,$$
$$d + p \rightarrow {}^3\text{He} + \gamma + 5.5\,\text{MeV} \quad ,$$
$${}^3\text{He} + {}^3\text{He} \rightarrow \alpha + 2p + 12.9\,\text{MeV} \quad . \tag{2.23}$$

As a result, four protons convert into an α-particle. The energy generated by this process is about 20 MeV. The total energy of the fusion reaction is what maintains the luminosity of the Sun.

Although the thermonuclear model successfully solved the "energy crisis," it had a difficult time in the process of its development. It is very instructive to consider one of these difficult situations. As has been mentioned in Sect. 2.7, the simplest variants of nucleosynthesis (combining a nucleon with an α-particle or fusion of two α-particles) are ineffective, since they give rise to unstable isotopes.

In the mid-sixties, the interesting idea was propounded that the problem can be resolved by assuming heavier elements to be formed by three-particle fusion, namely by transformation of three α-particles into a carbon nucleus. The idea of triple fusion is quite nontrivial. First, it implies a comparatively high concentration of α-particles. This item has been resolved on the basis of the model of cosmic helium nucleosynthesis (cf. Sect. 2.7). Another problem proved to be more complex: triple fusion can only be reasonably effective if the energy (or mass) of the elements on both sides of the symbolic equation for the reaction is the same:

$$3m_\alpha = m_\text{C} \quad .$$

In reality, a different relation,

$$3m_\alpha = m_\text{C} + 7.7\,\text{MeV} \quad ,$$

is fulfilled, however.

Consequently, a prerequisite for the effectiveness of triple fusion is the existence of the excitation level of the ${}^{12}\text{C}$ nucleus with the energy of

---

[12] It should be emphasized that owing to the low luminosity of the red dwarfs, gravitational energy is sufficient to provide for their existence for cosmological times.

[13] This cycle plays an especially important role, for the initial matter within stars consists primarily of protons.

7.7 MeV. At the time the triple-fusion hypothesis was suggested, an excitation level of the ^{12}C nucleus was unknown. With their back to the wall, the astrophysicists insisted on its existence, while nuclear physicists disagreed. Later on it was demonstrated by experimental investigation with accelerators that the ^{12}C nucleus does possess an excitation level with the energy of 7.7 MeV, i.e., the level that enables triple fusion to occur effectively. The problem of the formation of complex elements was thus resolved.

Physicists were unaccustomed to the feedback effect of astronomy on the progress in fundamental physics. That is why this fact strongly impressed the contemporaries. For example, R. Feynman devoted many a fervid line in his book "The Character of Physical Law"[14] to this event.

Stellar Collapse

But let us return to our main subject, stellar evolution. In the thermonuclear model of the structure of stars, energy liberation occurs by virtue of the fusion of light elements to form heavier elements. This synthesis is irreversible: heavy elements practically do not fragment into lighter ones. This irreversibility determines an inevitable process of star aging. As the atomic number of the elements participating in thermonuclear reactions increases, the energy liberated in such processes decreases. Finally, as the atomic number of the elements involved in the reactions approaches that of the elements of the iron group (Ni, Fe), fusion stops yielding energy dividends altogether. The star starts cooling down, and equality (2.19) providing for the equilibrium of the star does not hold any longer. The star then contracts very rapidly under the action of gravitational forces. Sometimes the collapse is accompanied by a beautiful firework display: the burst of a supernova. The luminosity of a supernova in the active phase exceeds that of a whole galaxy. However, not many astronomers have been lucky enough to observe such a sight: in our Galaxy they happen only every 20–30 years or so. Nowadays, owing to the advances in the techniques used in astronomy, it is possible to observe the bursts of supernova in other galaxies as well.

We are interested, however, in a less spectacular, but perhaps physically more important aspect of evolution, namely, the fate of a collapsing star. According to calculations, this fate is predetermined by the initial mass of the star, M_s (cf. Sect. 3.1.2). If $M_s < 1.4 M_{\odot}$, then, as a result of collapse, the star becomes a white dwarf. If $1.4 M_{\odot} < M_s < 5 M_{\odot}$, then it becomes a neutron star. For $M_s > 5 M_{\odot}$, the collapse goes on ad infinitum: the star is to become a black hole. The principal predictions of the thermonuclear model of stars is in fair agreement with observational data.

---

[14] R. Feynman: "The Character of Physical Law" (Cox and Wyman, London 1965).

It would be inappropriate, however, to consider this theory as proven. Theorists had many troubles with the observations of the neutrino radiation of the Sun which proved to be much smaller than calculated. They had to reconsider the parameters of the theory, in order to make the theoretical conclusions consistent with the observational data.

There is still another problem: the lack of convincing proof of the existence of black holes, although some indirect methods to observe them have been proposed. Whether this lack of proof is a drawback of the theory of stellar evolution or a consequence of the ineffectiveness of the existing techniques for black hole verification remains unclear.

3. The Universe and the Elementary Particles

3.1 On the Relation between the Characteristics of Stars and of the Elementary Particles

It has already been mentioned (cf. Sect. 2.9) that stars are giant spheres consisting primarily of protons and electrons with some admixture of α-particles ($\sim 6\%$) and, to a still lower extent, of nuclei of heavier elements. In real stars, the concentration n_s of the particles as well as the density of matter ϱ_s increases from the surface of a star toward its center.

In our analysis, the evaluation of the major characteristics of a star, the mass M_s and the radius R_s will be simplified to some degree. It will be assumed that the distribution of the density ϱ_s (or of the concentration, n_s) in a star is uniform, and that stars consist of protons and electrons only. This approximation reflects stellar evolution fairly well.

In the exact theory, both the nonuniformity of the density distribution and the complex chemical composition are taken into account. However, such calculations require cumbersome numerical computations which in fact obscure the transparent physical picture of stellar structure.

3.1.1 Stars and Planets: A Distinction

Prior to concrete calculations, let us try to make a distinction between the definitions of stars and planets. Like the stars of the main sequence, planets consist of nucleons and electrons but, as distinct from stars, in planets the nucleons and electrons are bound into neutral atoms. This physical difference is caused by the smallness of planetary mass, M_p, as compared to M_s. For instance, the mass of Jupiter − the largest of the known planets − is about equal to 10^{30} g, i.e., it is smaller than the solar mass by a factor of 1000. But even in Jupiter, a significant fraction of particles is bound in atoms, while the substance of a typical planet, the Earth, almost entirely (to $\sim 99\%$) consists of atoms.

The physical difference in the structure of stars and planets leads to an important difference in the principal visual characteristics, the luminosity. Planets are practically not self-radiating. There are two causes for this: first, their natural (intrinsic) temperature is low and, second, matter consisting of the plasma (protons and electrons) radiates much more strongly than an aggregate of neutral atoms. Accordingly, a "demarcation line" between the stars and the planets can be drawn on the basis of the state of the matter they contain. If the matter is ionized, then a compact celestial body is a star; if the matter consists of neutral atoms, then the body is a planet.

It should be stressed in this connection that up to now, only the planets of the solar system have been observed. However, one would hardly doubt the existence of planetary systems in the vicinity of other typical stars as well. That such planetary systems have not been revealed as yet is simply explained by the insufficient resolution of the available telescopes, which are unable to detect these practically dark objects in the bright field of a star.

3.1.2 Stellar Parameters: A Quantitative Evaluation

After these preliminary considerations, we can proceed to a quantitative evaluation of the principal stellar parameters M_s and T_s in terms of the gravitational interaction constant α_g and the masses m_p and m_e of the stable elementary particles of which the stars are made up.[1] The analysis will be based on the fundamental equilibrium condition expressing the equality of the potential energy and the kinetic energy associated with a pair of particles p and e^-,

$$\frac{GM_s m_p}{R_s} \sim \varepsilon_k \quad , \tag{3.1}$$

where ε_k is the average total energy of the proton-electron pair.[2]

In the considerations to follow, numerical factors of the order of unity will be omitted. This approximation corresponds to the physical accuracy of our estimates (cf. the assumption of uniformity of a star; see p. 61). From (3.1) and the obvious relations $M_s \sim \varrho_s R_s^3$ and $\varrho_s \sim m_p r_0^{-3}$, where r_0 is the average distance between the particles, the basic expression determining the radius R_s then follows:

$$R_s \sim \left(\frac{\varepsilon_k r_0^3}{G m_p^2} \right)^{1/2} \quad . \tag{3.2}$$

---

[1]   The neutron stars consist mainly of neutrons. In this section, the mass difference $\Delta m_N = m_n - m_p$ will be neglected as compared to m_p.

[2]   A similar condition has been repeatedly used in Chap. 2

Offhand, such calculations do not seem to bring us closer to our final goal, for the parameter R_s has been expressed in terms of new, unknown quantities, r_0 and ε_k. This conclusion is incorrect, however, because ε_k and r_0 are uniquely interrelated, while the quantity r_0 has only three preferred values corresponding to three classes of celestial objects: big planets/small stars, white dwarfs, and neutron stars. It is this possibility of finding three physically preferred states of celestial bodies which underlies the idea of the evaluation of the quantities M_s and R_s.

We consider first the relationship between the quantities r_0 and ε_k. This relationship is just a reflection of the uncertainty principle, the fundamental principle of quantum mechanics. Quantitatively, the uncertainty principle has the form $p \simeq \hbar/r$, where p and r are the characteristic values of the momentum and the size of a microsystem. In the case under consideration, this relation reads

$$\varepsilon_k \sim \frac{p^2}{m} \sim \frac{\hbar^2}{r_0^2 m} \quad , \tag{3.3}$$

where m is the characteristic particle mass which assumes two values, $m = m_e$ or $m = m_p$.

Now, let us consider possible values of the characteristic distance between the particles, r_0. As already mentioned, in microphysics there are three preferred values of particle spacing. The first one is equal to the atomic size. The size of light atoms of which planets consist has the order of magnitude of the size of the hydrogen atom,

$$r_{01} \sim r_H \sim \alpha_e^{-1} \frac{\hbar}{m_e c} \sim 10^{-8} \text{ cm} \quad . \tag{3.4}$$

The second value, r_{02}, is related to the quantum mechanical dimensions of an electron. As is well known, quantum mechanical effects become essential at distances between the particles comparable with the Compton length, $r \sim \hbar/mc$. In this case, the Pauli principle, which forbids a fermion to enter a region occupied by an identical particle, applies. In other words, at $r < \hbar/mc$, the Pauli principle brings about forces opposing further contraction. Thus, there is a characteristic distance,

$$r_{02} \sim \frac{\hbar}{m_e c} \sim 10^{-11} \text{ cm} \quad . \tag{3.5}$$

The third characteristic length, r_{03}, which is associated with the nuclear nucleon-nucleon interaction, is somewhat more involved. As is well known, the action radius of nuclear forces is $r_N \sim \hbar/m_\pi c$ (where m_π is the pion mass). There is, however, another characteristic length of nuclear interaction, $r'_N \sim \hbar/m_p c$. Hence, one should set

$$r_{03} \sim \frac{\hbar}{m_p c} - \frac{\hbar}{m_\pi c} \sim 10^{-14} - 10^{-13} \, \text{cm} \quad . \tag{3.6}$$

To each value of r_{0i} corresponds a definite value of the characteristic mass m. It is clear that the dimensions of the hydrogen atom and the Compton wavelength of the electron are determined by the quantity m_e, while the distance r_{03} is determined by the quantity m_p or m_π. Substituting these quantities and the respective expressions for r_{0i} into (3.2) yields three corresponding values of R_s:

$$R_{s1} \sim \left(\frac{\alpha_g}{\alpha_e}\right)^{-1/2} r_H \sim 10^{10} \, \text{cm} \quad , \tag{3.7}$$

$$R_{s2} \sim \alpha_g^{-1/2} \frac{\hbar}{m_e c} \sim 10^8 \, \text{cm} \quad , \tag{3.8}$$

and

$$R_{g3} \sim \alpha_g^{-1/2} \frac{\hbar}{m_p c} - \alpha_g^{-1/2} \frac{\hbar}{m_\pi c} \left(\frac{m_p}{m_\pi}\right)^{1/2} \sim 10^5 - 10^6 \, \text{cm} \quad . \tag{3.9}$$

Making use of the relation $M_s \sim \varrho_s R_s^3$ again, three characteristic values of the mass M_s are readily obtained:

$$M_{s1} \sim \left(\frac{\alpha_g}{\alpha_e}\right)^{-3/2} m_p \sim 10^{30} \, \text{g} \quad , \tag{3.10}$$

$$M_{s2} \sim \alpha_g^{-3/2} m_p \sim 4 \times 10^{33} \, \text{g} \quad , \tag{3.11}$$

and

$$M_{s3} \sim \alpha_g^{-3/2} m_p - \left(\frac{m_p}{m_\pi}\right) \alpha_g^{-3/2} m_p \sim 4 \times 10^{33} - 6 \times 10^{34} \text{g} \quad . \tag{3.12}$$

The obtained three pairs of characteristic values of the radius R_s and mass M_s can easily be brought into correspondence with the observational data. The pair M_{s1} corresponds to the largest planets or the smallest (by mass) stars, while the pairs M_{s2}, R_{s2} and M_{s3}, R_{s3} correspond to the white dwarfs and the neutron stars, respectively. It should be noted that more precise calculations yield 3×10^{33} g as the upper bound for the white dwarf mass and the range of $3 \times 10^{33} - 10^{34}$ g for the mass of a neutron star.

Up to now, we have not treated the most numerous stars – the thermonuclear stars of the main sequence. From the theory of thermonuclear reactions it follows that the fusion of two protons is most effective if the temperature T_0 approaches the value

$$T_0 \sim \frac{\alpha_e^2 m_p c^2}{k} \ . \tag{3.13}$$

More precisely, thermonuclear reactions begin to occur effectively when $T_0 \sim 10^7 - 10^8$ K. Utilizing (3.13) and making a few additional assumptions, simple relations can be derived for the averaged characteristics of thermonuclear stars. Similar approximations also follow from different physical considerations (e.g., from equating the thermal energy density and the radiation energy density in stars). The radii and masses of thermonuclear stars are given by

$$R_{s4} \sim \alpha_g^{-1/2} r_H \sim 10^{11} \text{ cm} \ , \quad M_{s4} \sim \alpha_g^{-3/2} m_p \sim 4 \times 10^{33} \text{ g} \ . \tag{3.14}$$

These values are in good agreement with the solar parameters (cf. Table 2.3).

It should be emphasized that the relations (3.14) represent the average parameters of thermonuclear stars. For example, the mass M_{s4} may vary within three orders of magnitude. Two questions remain, however.

i) *What class of objects corresponds to the mass range* $10^{30} - 10^{33}$ g? As already mentioned, the mass of 10^{30} g corresponds to the largest planet, Jupiter. According to the exact calculations, thermonuclear reactions occur in stars with mass $M_s > (2 - 3) \times 10^{32}$ g. The observed stars have a mass above $\sim 10^{32}$ g. The mass range of $M_s \sim 10^{32} - 3 \times 10^{32}$ is associated with the red dwarfs.

The mass range of 10^{30} g $\lesssim M_s \lesssim 10^{32}$ g is a terra incognita; no stars are observed in this range (cf. Sect. 4.12.1). This does not mean that there are no such stars in the Universe. On the contrary, it would be natural to think that the distribution of celestial bodies with respect to mass should exhibit a continuous transition from planets to stars. For two reasons, stars with a small mass are hard to observe, however: such stars would have a low luminosity and a relatively low temperature corresponding to the infrared range. Still, revealing and investigating stars with a low mass is a very tempting goal, for it is one of the means of solving the challenging problem of the missing mass (cf. Sect. 2.8).

ii) *What is the destiny of thermonuclear stars of very large mass,* $M_s \gtrsim 60 M_\odot \sim 30 \alpha_g^{-3/2} m_p \sim 10^{35}$ g.? It turns out that the star mass has a very sharp upper limit; for very large mass, stars become unstable in the following manner. Suppose a star has, for some reason, contracted somewhat. This will cause a temperature increase inside the star and hence an enhancement of thermonuclear reactions, thus leading to increased energy liberation which is accompanied by expansion. That is to say, radial oscillations will arise in the star. Calculations show that for $M_s \gtrsim 30\alpha_g^{-3/2} m_p$, these oscillations will grow so that the star will be either torn or stripped of its outer shells. For $M_s < 30\alpha_g^{-3/2} m_p$, the oscillations will be damped, i.e., the star will be stable. Observations do show that there are no stars with mass $M_s > 30\alpha_g^{-3/2} m_p$.

Thus, $M_s \sim \alpha_g^{-3/2} m_p$ gives a typical order of magnitude of the stellar mass. In a way, one may consider $M_s \sim \alpha_g^{-3/2} m_p$ as the mass of a "gravitational element." The enormous magnitude of this mass is caused by the weakness of the gravitational interaction.

Let us give without proof a useful expression for the average luminosity of a thermonuclear star:

$$L_s \sim \frac{(m_e c^2)^2}{\hbar} \alpha_g^{-1/2} \sim 10^{36} \, \text{erg s}^{-1} \quad . \tag{3.15}$$

From this, the average lifetime of the star is easily obtained:

$$t_s \sim \frac{a \alpha_g^{-3/2} m_p c^2}{L_s} \sim a \alpha_g^{-1} \frac{m_p}{m_e} \frac{\hbar}{m_e c^2} \sim 10^{18} \, \text{s} \quad , \tag{3.16}$$

where $a \sim 10^{-2}$ is the fraction of the stellar mass converted into energy in the process of thermonuclear reactions.

3.2 Structure of the Universe and the Mass of the Elementary Particles

In this section, and in the following ones, the instability of the structure of the Universe with respect to the numerical values of the fundamental quantities will be proven. This is to be understood in the sense that a small change in the fundamental constants results in a quantitative change of the structure of the Universe. This change consists in the disappearance of one or several basic elements of the Universe: the nuclei, the atoms, the stars, or the galaxies. The proof of the instability of the structure of the Universe is based on a standard method, the analysis of the behavior of functional dependence upon a small change of the argument of the function. However, the proposed procedure is somewhat unusual, and some effort is needed to get accustomed to it. The peculiarity of the situation lies in the object of the investigation, the Universe, as well as in the arguments with respect to which stability is studied, viz., the fundamental constants. In fact, our Universe is unique; it is the only specimen we are familiar with; thus, even imaginary manipulation of this unique object causes natural anxiety.

On the other hand, as a rule, the fundamental constants appear in physical experiments as fixed numbers, measured in numerous tests and independent of any factors. These qualities have earned the numbers the "honorable rank" of fundamental constants. At first glance, varying them would appear blasphemous; this is, however, the only way of investigating the stability of the structure of the Universe. Such treatment breaks established traditions; but progress in science results, to a great extent, from challenging traditions. However, breaking the accepted rules of the game does not always lead to true progress in science, of course.

In this section, the stability of the Universe with respect to the mass of the elementary particles will be analysed. It should be remarked here that we shall deal only with particles of which the major structural elements of the Universe consist, namely, the nuclei, the atoms, the stars, and the galaxies; i.e., we shall investigate the stability of the Universe with respect to three constants: the electron mass m_e, the proton mass m_p, and the neutron mass m_n. Let us start with the electron mass.

3.2.1 The Mass of the Electron

In its ground state, hydrogen is an absolutely stable element.[3] However, the margin of stability is not big; the hydrogen atom can collapse via

$$p + e^- \to n + \nu \quad . \tag{3.17}$$

This reaction, caused by the weak interaction, is already familiar to us from Sect. 2.7. In nucleosynthesis during the early stages of expansion, the temperature is sufficiently high ($kT_u \gtrsim m_e c^2$, $T_u \sim 10^{10}$ K), and reaction (3.17) occurs with a high efficiency. However, as the cosmological time t_u increases, the temperature T_u decreases, and in the epoch of neutral hydrogen formation when $T_u \sim 10^4$ K (cf. Sect. 2.8.3), reaction (3.17) is strictly forbidden because of the most rigorous physical law, the law of conservation of energy (mass): $m_e \sim 0.5\,\mathrm{MeV} < \Delta m_N = m_n - m_p \sim 1.3\,\mathrm{MeV}$. It is the inequality

$$m_e < \Delta m_N \tag{3.18}$$

which is a warrant for the absolute stability of hydrogen.

In analyzing the stability of the hydrogen atom, we shall continue to follow the method outlined above, viz. the hypothetical variation of the

---

[3] Here we exclude consideration of the hypothetical proton decay, the proton lifetime exceeding the age of the Universe by at least 20 orders of magnitude.

electron mass m_e. For the inequality sign in (3.18) to be reversed to yield $m'_e > \Delta m_N$, it is sufficient to have $m'_e \gtrsim 3m_e$. (Here the prime refers to the changed values of the fundamental constants.) With the aid of the standard formulae of quantum mechanics, the lifetime of the "hydrogen" atom can be calculated for the changed electron mass, m'_e. Thus, for $m'_e = 3m_e$, the time t_H during which the hydrogen atom would exist in the neutral state is about one month; for $m'_e = 4m_e$, it amounts to about 24 hours; t_H decreases very rapidly with further increase in the magnitude of m'_e.

What would happen to the Universe if the hydrogen atom were unstable? At first glance, it appears that the consequences would be significant but not catastrophic: all hydrogen-containing molecules would practically disappear. This optimistic expectation is absolutely incorrect, however: an increase in the mass m_e would lead to a major change in the entire structure of the Universe. If reaction (3.17) occurred in the Universe, the stars and the galaxies would change their appearance completely. It has been mentioned in Sect. 2.8.3 that the existence of neutral hydrogen is necessary for the formation of galaxies. The occurrence of reaction (3.17) would cause the neutral hydrogen being formed to convert to neutrons and neutrinos almost immediately. Galaxies and stars would then practically consist entirely of neutrons; complex forms of matter, the atoms and molecules, would be practically absent. It is precisely this situation which we refer to as the instability of the structure of the Universe — with respect to the constant m_e in the case under consideration.

The next obvious question to arise is: Is an increase of m_e by a factor of 3 to 4 large or small? In physics, such a laconic inquiry is in fact meaningless. The smallness of a quantity has a meaning in a relative sense only: Small compared to what?

In the case being considered here, there are characteristic ratios of the mass m_e to the average mass $\langle m \rangle$ of the rest of the particles or to the mass m_μ of the lightest of them, the muon. It turns out that $m_\mu/m_e \sim 200$. Compared to this quantity, the increase of m_e by a factor of three is small. Moreover, the mass of almost all elementary particles ($\sim 98\,\%$) is $\sim 1\,\mathrm{GeV}$. Consequently, the ratio of the average mass $\langle m \rangle$ of the elementary particles to the electron mass is $\langle m \rangle/m_e \sim 1000$.

In Fig. 3.1, the mass ratios of the particles known to be stable under strong interactions[4] are presented. The particle masses are placed in the order of their increase: $m_1 = m_e$; $m_2 = m_\mu, \ldots, m_{12} = m_D$ (D denoting a heavy D-particle). The values of $\lg(m_{k+1}/m_k)$, shown as the ordinate correspond to the symbols of the ratios m_{k+1}/m_k located on the abscissa. The figure clearly illustrates a strikingly large magnitude of the ratio m_μ/m_e

---

[4]   The mass of only such particles can be measured with sufficient accuracy.

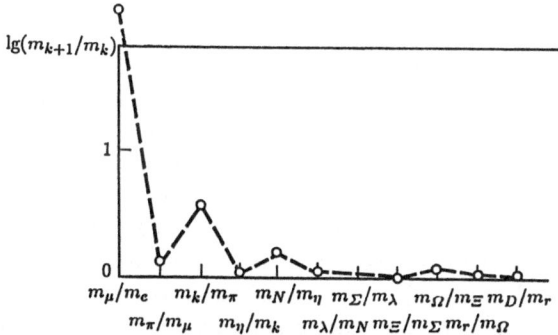

Fig. 3.1. The mass ratio of stable elementary particles. The marks on the abscissa symbolize the mass ratios of the adjoining (by mass) particles. The magnitude of these ratios is given, in the decimal logarithmic scale, by the ordinate

as compared to other values of the mass ratios. Below the horizontal line lies the "dangerous" range of the values of m_e' for which hydrogen atoms collapse if reaction (3.17) were to occur – with its further catastrophic consequences.

One remark in conclusion. From school physics on, we have been used to the notion of a relatively small value of m_e. In university physics texts, this smallness is usually associated with the leptonic character of the electron. Both leptons, the electron and the muon, are the lightest charged elementary particles, and they do not participate in the strong interaction. It is therefore tempting to simply explain the comparative smallness of the electron and muon mass by the fact that they do not take part in the strong interaction. This generally accepted interpretation was totally discarded, however, after the discovery (1977) of the τ-lepton – a particle not exhibiting strong interaction, yet twice as heavy as the proton.

The smallness of the value of m_e is a giant fluctuation which is, however, necessary for the existence of complex forms of matter. To explain in what sense the smallness of the mass m_e can be thought of as a fluctuation, we consider the following illustration. Suppose before Maxwell's discovery of the theoretical velocity distribution of molecules, the velocities of 100 molecules would have been within the interval 1000 ± 100 m s^{-1}, say, while for one molecule, it would have been equal to 1 m s^{-1}. Then the physicist conducting these measurements (not aware of Maxwell's law) would have been able to state with certainty that the velocity distribution of the molecules has a sharp peak at $v \sim 1000$ m s^{-1}, while the velocity $v \sim 1$ m s^{-1} of a single molecule is a large fluctuation.

3.2.2 The Mass of the Nucleon

It should be noted that an even stronger limitation on the upper bound for the varying mass m_e can be obtained from the condition for the occurrence of the thermonuclear reaction $p + p \rightarrow d + e^+ + \nu$ (cf. Sect. 2.9.2.1). This reaction is a basis for the existence of stars; it is caused by the weak interaction and is, therefore, very slow ($\sim 10^{10}$ years). Other reactions of the hydrogen-helium cycle are associated with the electromagnetic or the strong interaction, and their rate is very high (the characteristic times being of the order of several minutes). A star consisting, for example, of nucleons would explode. Consequently, the reaction of transformation of two protons into a deuteron determines the whole evolution of a star and, in particular, its prolonged existence. However, this reaction is only possible if a certain relation between the masses of the participating particles is fulfilled:

$$2m_p > m_p + m_n - \varepsilon_d + m_e$$

or

$$m_e < \varepsilon_d - \Delta m_N \sim 0.9 \, \text{MeV} \quad , \qquad (3.19)$$

where $\varepsilon_d \sim 2.2 \, \text{MeV}$ is the proton-neutron binding energy in the deuteron. The physical meaning of the above inequalities is quite simple: the reaction will occur if the total mass (or the corresponding energy) on the left-hand side of the equation (system prior to the reaction) exceeds the total mass (energy) after the reaction. The difference in energy before and after the reaction is equal to the sum of the neutrino energy and the kinetic energy of the particles generated in the reaction. Inequality (3.19) places a stricter limitation on the mass m'_e ($m'_e < 0.9 \, \text{MeV}$). It should be stressed that the admissible limit of the variation of m_e is understood as that at fixed values of Δm_N and ε_d. In principle, the possibility of all these quantities being interrelated cannot be ruled out; this possibility is considered below.

Let us pass on to the evaluation of the conditions imposed on the nucleon mass or, more precisely, on the difference $\Delta m_N = m_n - m_p$. In Sect. 2.7, helium nucleosynthesis was considered. There, it was implicitly assumed that the deuteron is a stable particle. If the deuteron spontaneously decayed, nucleosynthesis would not take place. Deuteron stability is a cornerstone of the whole consistent scheme of nucleosynthesis occurring in stars. But the margin of stability of the deuteron is not provided for well enough. A deuteron consists of a proton and a neutron, the latter decaying in the free state; in the bound state, within a deuteron, a neutron is stable, however. This stability owes to the fact that, besides the production of particles in a neutron decay, some energy (mass) has to be expended to destroy the deuteron. Deuteron stability implies the inequality $\Delta m_N < \varepsilon_d + m_e$. In conjunction with inequality (3.19), this yields the condition[5]

$$\Delta m_N < \varepsilon_d \quad . \tag{3.20}$$

This condition is indeed fulfilled in the Universe, but with a small margin. To prove this, we compare the difference Δm_N for the nucleons with the mass difference for isospin multiplets of other known particles (stable under strong interaction). Remember that an isospin family (multiplet) is a set of hadrons which are identical in all respects but in the electric charge and mass (the mass difference being small).[6] The nucleon (i.e., the proton and the neutron) is the best-known isospin doublet. The trinity of pions, $\pi^{\pm}, \pi^0$, is a typical isospin triplet.

Each isospin multiplet is characterized by the mass difference Δm of the particles constituting it. In Fig. 3.2, the mass differences Δm for stable isospin multiplets are shown. The region above the horizontal line corresponds to the values $\Delta m'_N$ for which the deuteron is unstable. On the basis of this figure, an interesting peculiarity can be visualized: the mass difference Δm_N for the nucleons is the smallest of all mass differences. Moreover, if the value of Δm_N were equal to the smallest of all other mass differences known, $\Delta m_{\Sigma} = m_{\Sigma^0} - m_{\Sigma^+} = 3\,\mathrm{MeV} > \varepsilon_d$, then the deuteron would be unstable.

Fig. 3.2. The mass difference of particles belonging to one isotopic multiplet. The symbols of the particles are given on the abscissa; the ordinate represents the numerical value of the mass difference

Hence, we encounter here an indication that the structure of the Universe depends on the extreme smallness of the magnitude of Δm_N as compared to other quantities Δm. In this sense, deuteron stability is a fluctuation. And finally: What is the structure of a universe with unstable

---

[5] More precisely, $\Delta m_N < \varepsilon_d - m_e \sim 1.7\,\mathrm{MeV}$.

[6] Regarding isospin, see Sect. 1.2.2.

deuterons? It is not difficult to sketch a scenario for such a universe: nucleosynthesis would not occur in it[7] and therefore, no complex elements would be present; it would consist of hydrogen, and thermonuclear synthesis would not take place; the stars not supplied with thermonuclear energy would rapidly collapse.

3.3 Structure of the Universe
and the Fundamental Interaction Constants

3.3.1 The Strong Interaction

It was mentioned in the beginning of this book that there is no "true" theory of strong (nuclear) interaction (cf. Sect. 1.1). The meaning of this statement is that no theory is available to describe the entirety of the phenomena caused by the nuclear forces with the help of one parameter, the interaction constant (like in quantum electrodynamics). To interpret the nucleon-nucleon interaction, one has to resort to introducing phenomenological parameters. Usually, for not too high velocities, i.e., in the nonrelativistic approximation, the main characteristic is the form of the interaction potential. An example of such a potential is relation (1.3) which provides a good description of the interaction between two nucleons when the spin and isospin effects are neglected. In the same approximation, the interaction potential can be represented by a square potential well with the depth $V > b\hbar^2/m_\mathrm{p}r_\mathrm{N}^2 \sim 25\,\mathrm{MeV}$. (Here, $r_\mathrm{N} = 10^{-13}\,\mathrm{cm}$ is the action radius of nuclear forces; $b \sim 1$).

As has been repeatedly mentioned, a stable bound state arises for $V = E_\mathrm{p} > E_\mathrm{k}$ where the kinetic energy of the nucleon is $E_\mathrm{k} \sim b\hbar^2/m_\mathrm{p}r_\mathrm{N}^2 \sim p^2/m_\mathrm{p}$. This relation is a consequence of the uncertainty principle mentioned above.

A deuteron is a system consisting of two nucleons, and all these considerations are applicable for it as well. Stability of the deuteron results from the condition $V_\mathrm{d} > 25\,\mathrm{MeV}$ which is fulfilled, but with a narrow margin. The binding energy ε_d for the nucleons in a deuteron is small ($\sim 2.2\,\mathrm{MeV}$). It cannot be decreased without altering the entire structure of the Universe, for this would lead to the reversal of the inequality signs in (3.19) and (3.20); these inequalities underlie the existing structure of the Universe. With the aid of (1.3), it is easy to determine the admissible limits of the decrease of α_s within which the inequality (3.20) still holds. By setting $V \propto \alpha_\mathrm{s}$, one finds that α_s cannot be diminished by more than 40 %.

---

[7] Let us mention another argument in favor of the absence of nucleosynthesis for growing Δm_N. The lifetime of the neutron then rapidly drops so that cosmic nucleosynthesis does not have enough time to occur (cf. Sect. 2.7 and Fig. 2.4).

More impressive is the upper limit on α_s, however. As a matter of fact, a stable biproton, ^{2}He (the system pp) "almost" exists in the Universe. As a result of electrostatic repulsion as well as of the effect of the Pauli principle, one has for the potential energy V_{pp} of the biproton: $V_{pp} < V_d$ and $V_{pp} < E_k$, with $V_{pp} - E_k \sim -0.05$ MeV. However, on increasing α_s by several percent, the biproton would become a stable particle; this would lead to severe changes in the structure of the Universe. Provided that the stable biproton does exist, the reaction

$$p + p \to {}^2\text{He} + \gamma \qquad (3.21)$$

is possible. As distinct from the main reaction of thermonuclear synthesis via deuteron formation, which is associated with the weak interaction, reaction (3.21) is determined by the electromagnetic interaction. While the rate of the former reaction is slow, reaction (3.21) is very fast. In fact, it is so fast that all the hydrogen would burn out within the first few minutes of the expansion of the Universe; the Universe would thus be devoid of hydrogen. The predominant element in such a Universe is helium, the principal reaction in stars being the triple fusion reaction $3\alpha \to {}^{12}\text{C}$. However, the feasibility of this triple fusion in a hypothetical Universe with a slightly increased value of α_s is very doubtful. The explanation of this statement rests on the specifics of the structure of nuclear levels. As mentioned before (cf. Sect. 2.9.2.1), the triple fusion reaction owes its reasonable efficiency to the existence of the excitation level with the energy ~ 7.7 MeV providing for the resonance character of triple fusion. A slight variation of the constant α_s would, generally speaking, lead to a rearrangement in the system of nuclear levels. The level width ($\sim$ keV) is small compared to the distance between the levels ($\sim$ MeV). It is therefore quite improbable that the efficiency of triple fusion will be preserved if the constant α_s is slightly changed.

Thus, a small increase (by several percent) of the constant α_s would lead to a system consisting merely of helium. Consequently, neither the constant α_s nor the binding energy ε_d of a deuteron can be even slightly increased without radically changing the structure of the Universe. It should also be emphasized in this connection, though, that the smallness of the binding energy ε_d is a giant fluctuation in the world of atomic nuclei. For almost all stable nuclei, this quantitiy exceeds 8 MeV; only for the deuteron is the binding energy per nucleon ~ 1 MeV.

It should be stressed in conclusion that each of the inequalities providing for the existence of complex structures in the Universe has been analyzed independently. Now the question arises as to whether a complex structure similar to that of our Universe can be obtained by simultaneously varying the masses m_e, m_p, and m_n and the constant α_s. Considerations based on the hypothetical existence of this stable ^{2}He nucleus and on the shift of the levels of ^{12}C nuclei prohibit significantly increasing the constants m_e, Δm_N and α_s. Some decrease of m_e, Δm_N, and α_s cannot be ruled out a priori, however. There are no straightforward and absolutely convincing arguments against such a possibility.

One consideration speaks against this hypothesis, however. It has been emphasized above that the magnitudes of m_e, Δm_N, and ε_d are extremely small in comparison with the average values $\langle m \rangle, \langle \Delta m \rangle$, and $\langle \varepsilon \rangle (\langle \varepsilon \rangle \approx 8\,\text{MeV}$ is the average binding energy of nucleons in a nucleus). We do not know the principles governing the formation of the fundamental constants. Experience suggests, however, that if there is a distribution of some quantity, and if this quantity takes, as a result of certain initial conditions, a value strongly differing from the average, then the probability of the occurrence of a still larger deviation is substantially decreased. The vulnerable point of this conjecture is that there is no clear physical definition of the initial conditions for the formation of the fundamental constants.

Although this matter can only be clarified by invoking additional hypotheses, the distributions over m, Δm, and ε themselves are based on well-known results of numerous investigations, and the fluctuational nature of the values m_e, Δm_N, and ε_d is indisputable. Some other considerations with regard to the existence of limits on the variation of these quantities toward lower values can also be presented in connection with certain constraints to be discussed below (cf. (3.31, 37)).

3.3.2 The Electromagnetic Interaction

There are several deliberations which impose limitations on the magnitude of the constant α_s. Most impressive is a possible constraint following from the instability of the proton (cf. Sect. 1.6). From (1.9) it follows that the proton lifetime t_p very strongly (exponentially) depends on α_e. The obvious limitation on the time t_p is the inequality

$$t_p > t_u \sim 1/H \sim 10^{17}\text{s} \quad . \tag{3.22}$$

By using (1.9), it is easy to find that this inequality is fulfilled if

$$\alpha_e < \frac{1}{80} \quad . \tag{3.23}$$

If the reversed inequality held, then practically all protons would eventually decay into positrons, photons, and neutrinos. Positrons and electrons would be annihilated, and it would be a dull and triste system. Consisting of massless particles no stable bound states would occur in it.

A lower bound for the constant α_e can be obtained by requiring that the grand unification be possible (cf. Sect. 1.5). In the framework of this unified theory, there are X-bosons with mass m_{wes} which also depends on α_s in an exponential manner. The opinion was expressed a long time ago (by M.A. Markov) that the largest possible mass of elementary particles is the Planck mass, $m_{Pl} \sim (\hbar c/G)^{1/2} = \alpha_g^{-1/2} m_p \sim 10^{19}$ GeV;[8] at present, this opinion is quite widespread. In the contemporary grand unification theory, the mass $m_{wes} \sim 10^{15}$ GeV is relatively close to the mass m_{Pl}. Thus, by combining the condition of the existence of the X-boson with the condition $m_{wes} < m_{Pl}$ (cf. (1.7) in Sect. 1.5.1), it is easy to obtain

$$\alpha_e > \frac{1}{170} \quad . \tag{3.24}$$

A constraint on the magnitude of α_e can also be obtained from nuclear physics. The electrostatic repulsion of protons in the nuclei should be smaller than their attraction caused by the strong interaction. As has been mentioned above, the binding energy of the nucleons in a nucleus is about 8 MeV, the electrostatic energy is $\varepsilon_e \sim e^2/r_N \sim \alpha_e m_\pi c^2$. Hence, the inequality

$$\alpha_e < \frac{8}{m_\pi}[\text{MeV}] \sim \frac{1}{20} \tag{3.25}$$

has to be fulfilled.

Another relation can be produced from the condition of feasibility of thermonuclear reactions in stars, (3.13). For thermonuclear reactions to be efficient, the relation $kT \sim \alpha_e^2 m_p c^2$ must be fulfilled; for stars, $kT \sim GM_s m_p/R_s$. Making use of expressions (3.14) for the mass M_s and the radius R_s of a star, an additional constraint is obtained:

$$\alpha_e \sim \frac{m_e}{m_p} \sim 10^{-3} \quad . \tag{3.26}$$

Hence, it follows from many considerations concerning the existence of stable bound states that in any universe where stable bound states do exist, $\alpha_e \sim 1/100$ has to hold approximately.

---

[8] This value of mass was first introduced by Planck early in this century.

3.3.3 The Weak Interaction

The weak interaction plays a decisive role in two aspects of the formation of the structure of the Universe. It determines: (i) the process of nucleosynthesis (cf. Sect. 2.7), and (ii) the principal thermonuclear reaction of the hydrogen-helium cycle in stars (cf. Sect. 2.9.2.1)

Let us consider the effect of the magnitude of the constant α_w on both processes. For the existing structure of the Universe, the following fact is of very great importance: the fraction of the α-particles (relative to the protons) is not zero or unity, but rather has an intermediate value (0.25 by mass). If α-particles were completely absent, the triple fusion $3\alpha \rightarrow {}^{12}C$ would not be possible and consequently, complex elements would not occur. On the contrary, if the protons were completely transformed into α-particles, hydrogen − the basis of complex organic compounds − would not occur.

But the fraction n_α/n_p is determined by the ratio n_n/n_p during the nucleosynthesis epoch ($t_u \sim 10$–200 s). The neutron concentration, in turn, depends on the neutron lifetime τ_n. It is known that $\tau_n \propto \alpha_w^{-2}$. Therefore, for the coupling constant of the weak interaction $\alpha'_w \sim 10\alpha_w$, say, the neutron lifetime would be small (~ 10 s) as compared with the duration of the nucleosynthesis epoch, so that the neutron concentration in that epoch would be negligible. Qualitatively, the dependence of the neutron concentration on t_u at a changed value of the constant $\alpha_w (\alpha'_w = 10\alpha_w)$ is shown as a dashed line in Fig. 2.4. In this case, there would be no neutrons and cosmic nucleosynthesis would not occur. The Universe would consist almost exclusively of hydrogen.

In the opposite case, i.e., at a significantly decreased constant $\alpha_w (\alpha''_w \lesssim 0.1\alpha_w)$, reactions (2.18), which determine the statistical equilibrium between neutrons and protons, set in too late, i.e., when the nucleosynthesis era is already over. At a decreased value of α_w, reactions (2.18) would be so slow during nucleosynthesis that statistical equilibrium would not be provided for. The relative fraction of the neutrons will therefore not vary after the epoch of high temperatures of the Universe ($T_u \sim 10^{13}$ K, the hadronic era), when the ratio $n_n/n_p \sim 1$. This case is represented in Fig. 2.4 by the dot-and-dash line; here, practically all protons will eventually transform to α-particles.

A significant change of the constant α_w would undoubtedly lead to an essential change in the character of the evolution of the stars of the main sequence. Unfortunately, the corresponding calculations are lacking, and one has to resort to qualitative considerations. A substantial increase of the constant α_w would result in an increased rate of the principal thermonuclear reaction by which two protons transform to a deuteron (cf. Sect. 2.9). At a sufficiently high rate of this reaction, the heat liberated would be not completely transferred outwards; the star interior would be heated; this heating would, in turn, lead to further enhancement of thermonuclear reactions and, ultimately, to an explosion of the star.

Conversely, a significant decrease of the constant α_w would lead to a rapid transformation of the main sequence stars into white dwarfs and neutron stars. Indeed, the only conceivable way of compensating for the lowering of the energy release associated with thermonuclear reactions caused by a decrease of the constant α_w is to increase the star temperature. At an invariable mass M_s, this is only possible by decreasing the star radius R_s. As R_s approaches the size of a white dwarf, quantum effects come into play (cf. the Pauli principle, Sect. 1.3), opposing further contraction of the star.

3.3.4 The Gravitational Interaction

Strictly speaking, variation of the gravitational interaction – at least within 1–2 orders of magnitude – does not lead to any qualitative change of the physical structure of the Universe. The gravitational constant determines the characteristics of stars, galaxies, and perhaps the Universe as a whole (see below). It seems that a relatively small variation of this constant only results in corresponding changes in these structural elements, the physical foundations of their existence not being violated.

This conclusion is correct, assuming the openness of the Universe, however. Should the Universe be closed, i.e., for $\varrho > \varrho_c$ (cf. Sect. 2.4), certain constraints on the possible variation of the constant α_g will be imposed. These constraints follow from conditions (2.21–22), necessary for the formation of galaxies. Although, as emphasized in Sect. 2.8, there is no proven theory of galaxy formation, the conditions under which galaxies can be generated can be formulated. In particular, conditions (2.21–22) containing the cosmic background radiation temperature t_u must be satisfied. Both conditions imply that in the process of its evolution, the Universe had to go through an epoch during which the temperature T_u was below a certain value $T_u^{(m)}$.[9]

In an open universe the temperature T_u goes down continuously until it reaches the value $T_u = 0$ (at $t_u \to \infty$). Therefore, in the life of an open universe, there always is an epoch during which the condition $T_u < T_u^{(m)}$, determined by relations (2.21–22) begins to hold.

By contrast, the situation in a closed universe is quite different. In this case, there always is a minimum temperature, $T_{u\,min}$, corresponding to $R_u = R_{u\,max}$. The temperature T_u will then increase as contraction proceeds. Hence, in a closed universe in which galaxies form, the condition

---

[9] $T_u^{(m)}$ designates the temperature for which conditions (2.21–22) are fulfilled.

$$T_{u\,min} < T_u^{(m)} \qquad (3.27)$$

has to be fulfilled. As an order-of-magnitude estimate, one has, for a closed universe, $t_{u\,max} \sim R_{u\,max}/c \sim GM_u/c^3$. This relation can be considered a consequence of: (i) dimensionality considerations, or (ii) an empirical relation obeyed by our Universe.

From the anthropic principle ($t_u \sim t_s$, cf. Sect. 3.6 below) as well as from the observational data, it follows that

$$M_u \sim \alpha_g^{-2} m_p \quad . \qquad (3.28)$$

Then $t_{u\,max} \sim G\alpha_g^{-2} m_p/c^3$, which yields, upon substitution into the expression for the temperature T_u, (2.13), the value of $T_{u\,min}$ for a closed universe,

$$T_{u\,min} \sim \frac{\alpha_g^{1/4} m_p c^2}{k} \quad . \qquad (3.29)$$

Substituting this expression into relations (2.21–22) and setting $E_\gamma \sim kT$, $\varepsilon_H \sim \alpha_e^2 m_e c^2$, we obtain the limitation on the magnitude of the constant α_g:

$$\alpha_g^{1/4} < S^{-1} \qquad (3.30)$$

(where $S = n_\gamma/n_p$; for the Universe, $S \sim 10^8 - 10^{10}$) and

$$\alpha_g^{1/4} < \alpha_e^2 \left(\frac{m_e}{m_p}\right) \quad . \qquad (3.31)$$

An additional limitation on the magnitude of α_g can be obtained, under the same assumptions, by using the condition $t_u < t_p$ (where t_p is the proton lifetime, cf. Sect. 1.6). With the aid of the formulae (1.9) and (3.28) one readily obtains

$$\ln \alpha_g \gtrsim - (\alpha_e)^{-1} \quad . \qquad (3.32)$$

It should be noted that for our Universe, (3.30–32) are satisfied, but with a narrow margin.

It is interesting to remark that a relation closely similar to (3.21) was obtained long ago by Landau and Pomeranchuk, who were guided by completely different considerations. They analyzed the self-consistency of quantum electrodynamics at very large momentum or mass transfer and came to a conclusion that at $m \gtrsim (\hbar c \alpha_e/G)^{1/2}$, quantum electrodynamics breaks down as a logically self-consistent theory. To remove the inconsistency, it is sufficient to assume that for the transferral of so large a mass, the net interaction is determined by the gravitational forces and that the relation $\ln \alpha_g \sim -(\alpha_e)^{-1}$ is fulfilled.

3.4 The Dimension of Space

Of all the fundamental constants, perhaps the best known or at least the most familiar one, is the dimension of physical space, $N = 3$. Offhand, the very inquiry about the nature of the dimension appears absurd: Would the author dare to tackle the Holy of Holies of physics, the dimension of space, and use the method of variation of the fundamental constants to answer the question of why we live in three-dimensional space? This rhetorical question can be quite reasonably justified. Variation of such a fundamental characteristic as the dimension N may lead to unpredictable changes of physical laws; thus, one has to assume full responsibility for such a step. The author does venture to change the dimension of space under the premise that it is possible (and even not difficult) to predict some physical laws upon changing the dimension of space, provided that another property, viz., its being Euclidean, i.e., uniform and isotropic, is preserved. Since such a statement might appear presumptuous on the part of the author, it is advantageous to refer to the authority of Ehrenfest, whom we shall follow in this section. Indeed, it was Ehrenfest who, in 1917, was trying to answer the question of why physical space is three-dimensional.

To approach the answer to this question, it is perhaps best to regress. From school physics, we are familiar with the analogy between Coulomb's law and the law of universal gravitation (Newton's law). In both cases, the force is $F \propto r^{-2}$. In textbooks on physics (in particular, in the university courses) these laws are treated separately. This lack of coherency obscures a profound relationship of the electromagnetic and the gravitational forces with the properties of space, in particular, with its dimension.

Two properties are common to the gravitational and the electromagnetic interactions: both are weak and long-range. In modern language, this means that the mass of the exchange particles is zero in both cases, implying that the interaction radius is infinite and that the interaction constants are small, $\alpha_g, \alpha_e \ll 1$ (cf. Sect. 1.4 and Tables 1.2 and 1.3).

In the more familiar language of Faraday-Maxwell physics, these properties mean that the lines of force, originating at the point where their source is located, run to infinity, not intersecting with each other, provided that no other source is present. The fact that the lines of force extend to infinity reflects the long-range character of the gravitational and electromagnetic forces; the absence of intercepts signifies that there is no reciprocal action between the lines of force, i.e., that the interactions under consideration are weak.

The combination of both properties, the weakness and the long-range action, is not characteristic of other interactions. The so-called weak interaction has a limited radius, the mass of the exchange particles being nonzero. The mass of the exchange particles of the strong interaction, the gluons, is zero. However, the constant (at $r \sim r_N$) is $\alpha_s \sim 1$, and therefore the lines of force are essentially intersecting to form new lines of force; this reflects the existence of the color charge of gluons and their strong mutual interaction.

Let us, however, return to the long-range interactions. As the lines of force from a point source do not cross and extend to infinity in this case, their total number remains constant in space. This number is proportional to the charge (mass) of the particle.

The force F exerted by one particle on another particle, a distance r apart, is proportional to the density n_i of the lines of force. Accordingly,

$$F \propto n_i = \frac{f}{4\pi r^2}.$$ (3.33)

The proportionality constant f in (3.33) is by definition equal to the product of the charges of both particles (Coulomb's law) or of their masses (Newton's law). The denominator in (3.33) gives the surface area S of a sphere of radius r. For three-dimensional space, this quantity equals $S_3 = 4\pi r^2 = a_3 r^{3-1}$.

These considerations can be repeated for the more general case of an N-dimensional space. The surface area of a sphere in such a space is $S_N = a_N r^{N-1}$. Therefore, the force F_N, acting in an N-dimensional space, is

$$F_N = \frac{b_N}{r^{N-1}} .$$ (3.34)

Accordingly, the potential energy has the form

$$U_N = \frac{-b_N}{[(N-2)r^{N-2}]}$$ (3.35)

where $N \neq 2$; for $N = 2$, the dependence is logarithmic. It should be stressed once again that these expressions apply for integer N, but only for long-range forces in the quasistatic approximation, i.e., for motion in a central force field. The basic statement is as follows: the existence of stable orbits in a central force field in an N-dimensional space is determined by (3.34) and consequently by the dimension N.

From mechanics, it is known that the existence of stable orbits depends on the form of the r-dependence of the effective potential

$$U_{Ne} = U_N + \frac{M^2}{2mr^2}$$

(where $M^2/(2mr^2)$ is the centrifugal energy, M the angular momentum and m the mass of the body moving at a distance r). A stable state is possible if the dependence $U_{Ne}(r)$ has a minimum at a value of r different from zero or infinity.[10]

Let us present the results of an analysis of the function $U_{Ne}(r)$ with regard to the existence of an extremum:

1. For $N>4$, the dependence $U_{Ne}(r)$ has a maximum at $r\neq0$ and a minimum at $r = 0$ which corresponds to merging of the two particles.
2. For $N = 4$, the dependence $U_{Ne}(r)$ is given by a monotonically decreasing function exhibiting no extrema.
3. For $N = 2$ and $N = 3$, the function $U_{Ne}(r)$ has a minimum at $r\neq0$ and $r\neq\infty$.
4. For $N = 1$, the function $U_{Ne}(r)$ is monotonically increasing.

The existence of a minimum in the dependence $U_{Ne}(r)$ is a necessary condition for the stability of motion. For this reason, the existence and the properties of closed orbits are determined by the dimension for space, N. For $N\geq4$, there is no minimum at $r\neq0,\infty$; consequently there are no stable closed orbits. Any motion caused by long-range forces would be of one of the following two types: either it is infinite (the body escapes to infinity) or otherwise, the moving body falls on a massive central body. At $N = 2$ or $N = 3$, all types of motion are possible: infinite motion, fall onto a central body and, notably, motion in stable, closed orbits. For $N = 1$, only finite motion is possible; a body cannot escape to infinity.

Let us clarify the last statement. In the one-dimensional case, no orbital motion occurs, and the centrifugal potential is zero ($M = 0$). The effective potential, (3.35), is then $U_1 = b_1r$, and the force $F_1 = \text{const}$. This corresponds to an infinitely deep potential well. To remove the body from this well, an infinitely large force has to be applied; this means it would be impossible for the body to escape to infinity.

Hence, the degree of stability grows as the dimension N decreases. For $N\geq4$, there are no analogues to planetary systems. Similar considerations in the framework of quantum mechanics (F. Tangherlini; L. Gurevich and V. Mostepanenko) have demonstrated that for $N\geq4$, stable atomic systems do not exist, either. It appears that the absence of analogues of planets and atoms for $N\geq4$ is a clue to the understanding of the significance of the space dimension $N = 3$ (cf. also Sect. 3.6).

---

[10] Note that henceforth the attractive forces are considered for which $U_N<0$.

3.5 Structure of the Universe and Quantum Numbers of Elementary Particles

The existence of the fundamental quantum numbers is necessary for the existence of fundamental stable states.

The electron spin (or, more precisely, its half-integer value) leads to the Pauli principle that forbids electrons to occupy identical states. This principle is a cornerstone of the periodic system of the elements, i.e., of the diversity of chemical compounds (cf. Sect. 1.3). Without it, all atomic electrons would occupy the ground state; there would be no valence bonds associated with the structure of the Periodic Table and thus, complex compounds result could probably not be formed.

Even more significant is the obvious role played by the isospin. In particular, a decisive role is played in the structure of the Universe by the iso-doublet character of the nucleons, i.e., the fact that a nucleon multiplet includes both a proton and a neutron. Indeed, were the nucleons iso-singlets, obvious "catastrophic" consequences would follow. For example, if neutrons but no protons existed, then the charge of the nuclei would be zero, and there would be no atoms caused by the electromagnetic interaction. Conversely, if only protons, but no neutrons existed, then the fusion of two protons would not occur at all (hence, no deuteron!), or the reaction would be determined by the electromagnetic interaction, cf. (3.21). In this case, as has already been mentioned in Sect. 3.3.1, all available hydrogen would burn up in the process of nucleosynthesis. The more complex reactions of nucleosynthesis would occur via the electromagnetic or strong interaction, giving rise to the instability of stars.

Stability of atomic nuclei consisting of protons only would also be decreased; the physical foundation of this statement is evident. The presence of neutrons in the nuclei increases the distance between two neighboring protons and consequently reduces the influence of two factors which enhance destruction of nuclei: the Pauli principle and electrostatic repulsion. The isotriplet nature of the nucleons, i.e., the existence, in addition to the proton and the neutron, of a negatively charged analogue of the proton – a particle with a positive baryon charge and a negative electric charge – would give rise to (almost) electrically neutral nuclei, implying the absence of complex atoms. This conclusion follows from the laws of general dynamics: Any system tends to occupy the lowest energy state. In the case in question, this corresponds to a minimum of electrostatic energy, i.e., to mutual compensation of the charges of the protons and their "negative" analogues.

In the framework of the theory of grand unification, other quantum numbers also find a natural explanation, as does the existence of a relatively large number of leptons. It was mentioned in Sect. 2.6 that the baryonic asymmetry of the Universe is the basis of its complex structure. A most natural and simple interpretation of baryonic asymmetry, taking into account charge symmetry violation, is based on the approach developed by Kobayashi and Maskawa (cf. Sect. 2.6), requiring the existence of at least six different quarks and leptons. The quarks have to differ in their quantum numbers; for practical use, they are designated as strangeness, charm, etc. The Kobayashi-Maskawa approach might also give a key to the answer to the question that has preoccupied many generations of physicists: What is the purpose of muon? The answer would be that the muon is one of the leptons which provides for the baryonic asymmetry of the Universe.

Obviously, the Kobayashi-Maskawa interpretation described above is not a proof; it is, however, the simplest approach, although not the only one. The present situation with respect to quarks and leptons suggests once again that Nature does not tolerate "architectonic redundancies."

3.6 The Anthropic Principle (cf. Sect. 4.10)

Dicke appears to have been the first to formulate, in 1961, the anthropic principle. Later on, this principle was vigorously developed by the best British physicists and astrophysicists (S. Hawking, M. Rees, B. Carter, J. Barrow). It is very likely that the development of the anthropic principle was strongly influenced by Dirac, who worked at Cambridge for a long time (cf. Sect. 3.7).

Dicke's priority in formulating the anthropic principle is referred to here with some uncertainty because of the vagueness (from a physicist's viewpoint) with which the principle was expressed by the British researchers: The presence of an "observer" in the Universe places constraints on the physical laws. Usually, the notion of "observer" is interpreted in a broader sense than that referring to Man; rather, an "observer" is understood as a highly organized being. Nevertheless, the uncertainty in the formulation of this principle does not permit associating it with a particular inventor; there are reasons for attributing it to Ptolemy, for that matter.

But it is perhaps more purposeful to turn from discussing priority claims of the anthropic principle to its application. It will be demonstrated that contemporary understanding of the anthropic principle differs radically from its interpretation by the ancient astronomers.

It has been repeatedly mentioned (cf., e.g., Sect. 2.2) that the average density of matter in the Universe appears to be $\varrho \sim \varrho_c$. This coincidence has no strict physical corroboration; from general physical principles, the density ϱ could have any value. Still, that relation does (approximately) hold for our Universe. Its explanation is a "classic" example of the application of the anthropic principle.

If the inequality $\varrho \ll \varrho_c$ held in the Universe, then the inertial forces would outweigh the gravitational forces, and no galaxies (and hence no "observer") could be formed. Even with $\varrho \sim 0.1\varrho_c$, the theory of galaxy formation raises significant questions (cf. Sect. 2.8).

If the inequality $\varrho \gg \varrho_c$ were realized in the Universe, the theory of galaxy formation would cause no difficulties. However, the cycle time of a closed universe rapidly decreases with increasing ϱ. Therefore, in such a universe with $\varrho \gg \varrho_c$, there would not be enough time to create an "observer." For example, according to geological data, even the protozoans arose "only" 3 billion years ago.

Another example of the anthropic principle is the interpretation of the magnitude of the constant α_g, which is extremely small compared to unity. As indicated above (cf. Sect 3.1.2), the parameters of the stars have a weak dependence on the constant α_g. Therefore, on increasing the constant α_g even by several orders of magnitude, nothing terrible will happen to the stars: they will continue to exist, although their mass will be reduced significantly. Along with the reduction of mass, the lifetime t_s of a star will be reduced, however ($t_s \propto \alpha_g^{-1}$, cf. (3.16)). Thus, increasing α_g, e.g., by a factor of 100, will lead to the reduction of a star's lifetime t_s by the same factor. This circumstance will prevent an "observer" from being formed. Consequently, even a relatively small increase of α_g (by 1–2 orders, say) is not admissible in a Universe where an "observer" is present.

Let us consider another example. It has been mentioned earlier (Sect. 2.8) that the radiation of a system drops steeply if recombination, i.e., organization of protons and electrons into neutral atoms, occurs. This process is very efficient at $T \lesssim 1000$ K. As mentioned above, the surface temperature of visible stars $T_s \gtrsim 2000$ K (cf. Table 2.2). However, as has been pointed out by B. Carter, the temperature T_s very strongly depends on the ratio of the constants α_e and α_g. To render the recombination ineffective, the condition

$$\alpha_g^{1/2} \gtrsim \alpha_e^6 (m_e/m_p)^2 \tag{3.36}$$

is necessary. This condition is fulfilled for the Universe, but again, just barely. Were the reversed inequality fulfilled, the luminosity of stars would drop precipitously, and the advent of an "observer" on the planets – if at all possible – would be strongly impeded.

Our last example is a physical one. For decades, physicists and astrophysicists have been contemplating the amazing empirical relation between the atomic constants and Hubble's constant, which characterizes the Universe. This relationship can be expressed as follows:

$$1/H_0 \sim \frac{\hbar}{m_e c^2} \alpha_g^{-1} \quad . \tag{3.37}$$

Of course, this relation is fulfilled only roughly, with an order-of-magnitude accuracy only. However, taking into consideration that the constants entering this relation are significantly different in their values, it appears very remarkable. There have been many attempts to explain this relation. In particular, Dirac proposed his own interpretation, based on the hypothesis that the fundamental constants vary with time (see next section). Perhaps the most transparent explanation of this relation follows from the anthropic principle. A most favorable situation for the rise of an "observer" is when t_s, the lifetime of a star that does not exceed the age of the Universe t_u, coincides with t_u within one order of magnitude. Thus, an "observer" comes into being, provided that $t_u \sim t_s$. Using expressions (3.16) and (2.6) for t_s and t_u, respectively, yields relation (3.37). Setting $t_u \sim GM_u/c^3$ for a closed universe, one can obtain from this relation an expression for the mass of the Universe,

$$M_u \sim \alpha_g^{-2} m_p \tag{3.38}$$

(where the factor m_p/m_e has been omitted in view of the hugeness of M_u). Maybe it is an amazing freak of fortune that the latter relation is fulfilled for the observable Universe.

There may be various opinions about the physical significance of the anthropic principle, but it should be definitely acknowledged in one respect: it has essentially contributed to legitimizing the question: "Why is the world the way it is?"; previously, one was content to ask: "What is the world like?".

3.7 On the Numerical Values of the Fundamental Constants

Let us summarize what has just been said. The structure of the Universe is unstable with respect to the numerical values of the fundamental constants upon relatively small changes in the magnitude of the fundamental constants: four interaction constants, α_g, α_w, α_e, and α_s; the masses m_p, m_n, and m_e of the particles constituting the structural elements of the Universe; and the space dimension N. By qualitative change of the structure is meant the impossibility of the existence and/or rise in the Universe of one or several stable bound ground states: nuclei, atoms, stars, and galaxies. It is these structural elements which determine the diversiform appearance of our Universe. In its current cycle, our Universe is a complex system, as compared to those universes whose evolution can be described on the basis of the known dynamical laws, but with changed values of the fundamental constants.

Some questions may be raised here. In our deliberations, we have usually concentrated on varying one constant only; would our conclusions be altered if two or all of the constants are varied? The answer is probably no. The set of the constants m_e, Δm_N, α_e, and α_s constitute a closed system of interrelated constants, the number of the relations exceeding four. Thus, the range of their variation is limited. The only possibility of realizing complex universes is by reducing the mass parameters m_e and Δm_N. This possibility cannot be ruled out, although it seems quite improbable, owing to the fluctuational character of the values of m_e, Δm_N, and ε_d (cf. Sects. 3.2 and 3.3).

More difficult is another question, which at present cannot be answered unequivocally: Can universes as complex as ours (or more complex) occur if both the dynamical laws and the fundamental constants are varied? There is no tool available with which to derive the equations of quantum field theory in a unique way, proceeding from some "first principles." However, as exemplified by the analogy between Newton's law and the Coulomb approximation, it is feasible to deduce these laws by specifying the geometrical properties of space and making some subsidiary postulates. Having fixed the characteristics of space as well as the properties of the exchange particles, one can derive the classical equations of electrodynamics.

Thus, the dynamical equations are largely predetermined by geometry: to some extent (though, as we should like to stress, not unequivocally), they are determined by the properties of space that − at least for the weak, strong, and electromagnetic interactions − can be identified with the space in which the special theory of relativity holds, i.e., with Minkowski space. The choice of this space for the dynamical events to occur in is predetermined by its extreme simplicity: it is the only planar (Euclidean), maximally symmetrical (i.e., isotropic and homogeneous) space.

A different question is how many interactions (or types of exchange particles) occur in the Universe. The appearance of our Universe is determined by four interactions; all four actively participate in the formation of its complex structure. A reduction of the number of interactions changes the structure of the Universe qualitatively (cf. Sect. 3.3); increasing this number would be probably an "architectonic redundancy."

After these remarks, it would be worth-while to present a table summarizing possible limits on the variation of the fundamental constants. In Table 3.1, the symbols $f_+(f_-)$ denote the factors by which the numerical values of the fundamental constants pertinent to the existing Universe may be multiplied without breaking up all the principal bound ground states: nuclei,

Table 3.1. Possible limits on the variation of fundamental constants

Justification	f_-	Constant	f_+	Justification
Fluctuational character of the constant	?	m_e	2.5	Existence of atoms
Existence of atoms	0.4	Δm_N	1.6	Existence of nuclei
Feasibility of constructing a unified field theory	0.8	α_e	1.6	Stability of protons and nuclei
Existence of complex nuclei	0.9	α_s	1.1	Existence of complex nuclei; existence of hydrogen
Existence of hydrogen	0.1	α_w	10	Existence of complex elements
Fluctuational character of the constant	?	α_g	10^4	Formation of galaxies; the anthropic principle
The anthropic principle; the absence of analogues of planetary systems	1	N	1	The absence of analogues of planetary systems and atoms

atoms, stars, and galaxies. In the column "Justification" the considerations underlying the determination of the factors f_+ and f_- are briefly indicated; it should be noted that these considerations are not equally justified for the different cases.

Those estimates resting on the firmly established laws of nuclear and atomic physics can be considered most reliable. The argument based on proton decay is quite plausible but remains hypothetical.

The arguments based on the anthropic principle have, in a way, a qualitative character, at least at the present stage of our understanding of Nature. It should only be remarked that the fact that a space with the dimension $N = 1$ or $N = 2$ is unrealistic is connected with the impossibility of forming structures in it which are as complex as those possible in three-dimensional space. It should be mentioned that in the framework of the theory of general relativity, gravitational interaction in spaces with the dimension $N = 1$ or $N = 2$ is absent. The "fluctuational" argument contains hypothetical elements as well;[11] we only mention here that the fluctuational aspect of the

---

[11] This argumentation is elaborated on in the next section.

constant α_g is associated with the extreme smallness of its magnitude as compared to unity or other constants.

It should be emphasized in conclusion that the quantities determining the complex structure of the Universe can be reduced to dimensionless combinations, i.e., to the following numbers:

Constants	α_g	α_w	α_e	α_s	N	m_e/m_p	$\Delta m_N/m_p$;
Numbers	10^{-38}	10^{-11}	10^{-2}	1	3	10^{-3}	10^{-3} .

Note that since the constants α_g, α_w, and α_s are mass dependent, their numerical values have been taken for the characteristic values of the momenta or masses which pertain to the values of these constants under conditions in which they determine the principal structural units of the Universe. For instance, the weak interaction plays a decisive role in cosmological nucleosynthesis (cf. Sect. 2.7) at $kT \sim m_e c^2$. Consequently, in the expression for α_w, the value $m = m_e$ has been used. To evaluate the constants α_g and α_s, m has been set equal to m_p.

Thus, one can conclude that the structure of the Universe is determined by the above numbers. In a certain sense, this statement signifies a return to the concept of the ancient natural philosophers (Plato, Pythagoras), assigning to numbers a crucial role in the structure of the Universe. It has to be noted, however, that our current understanding of these numbers is much more profound than that of the ancient philosophers.

3.8 Conclusion

Hence, we are confronted by the instability of the structure of the Universe with respect to the numerical values of the fundamental constants. A comparatively small change in these values (cf. Table 3.1) leads to a universe significantly simplified compared to our Universe, i.e., to a universe from which the harmony of complex structures is missing. How should this fact be treated? It would be simplest to just overlook it; however, it is now practically impossible to do so. To substantiate this, we should digress from physics and formulate more precisely the basic notion we are going to deal with: What is the meaning of the word "Universe"?

3.8.1 Defining the Words "Universe" and "Metagalaxy"

Usually, this notion combines three essentially different meanings: (i) the Universe is everything that exists; (ii) the Universe is the observable region of space; and (iii) the Universe is the system which arose out of a (quasi) point some 20 billion years ago and further developed according to Friedmann's model.

At first glance, all three definitions appear to be consistent with each other. Indeed, the observable part of the Universe has the size $R_0 \sim 10^{28}$ cm; according to Friedmann's theory, $R_0 \sim c/H_0 \sim 10^{28}$ cm, as well. Furthermore, the visible Universe is simply "everything in the world." There is a vulnerable point in this reasoning, however: Why should all we observe be synonymous with all that exists? Why should the formation that arose some 20 billion years ago out of a point be "everything in the world"? These two questions may be reformulated: What was there before the Universe started expanding? What is there beyond its limits? Logically, the answer "nothing" is not contradictory.

However, the rise of an object like the Universe out of "nothing" contradicts our entire physical experience, although on a much smaller scale. Nevertheless, for a time, this "minor" difficulty was ignored. One reason is quite obvious: the ideas concerning the early Universe were too indefinite. The singularity point was far away, and one preferred not to touch upon such intricate questions. However, more recently, our understanding of the early Universe has made great progress. Nowadays one speaks with relative certainty about the processes which occurred at the time $t_u \sim 10^{-35}$ s (cf. Sect. 2.6), and it seems quite natural to ask what existed at $t_u = 0$ and $t_u < 0$. Any answer of the kind, "There was something", necessarily leads to the conclusion that the three definitions of the Universe are not consistent with each other; thus one has to sacrifice either the first definition or the last two.

Unfortunately, to date there is no consensus on the means of solving this terminological (and not only terminological) problem of paramount importance. The observable region of the Universe of radius $R_0 \sim 10^{28}$ cm is sometimes referred to as "the Universe" (as we have done in this book until now), sometimes as "the Metagalaxy", and sometimes as "the world"; occasionally, the terms "mini-Universe" or "part of the Universe" are used. Therefore, our definitions have to be given a more precise meaning. Although we are aware that there is some arbitrariness in this procedure, we shall reserve the term Universe for all being, i.e., for the whole world. The observable part of the Universe that originated, according to Friedmann, approximately 20 billion years ago, will be referred to as the Metagalaxy. It should be emphasized that by introducing these distinctions, we are departing from the terminology adopted earlier in this book. Up to now, there was no need for analyzing terminological nuances, however, and we were using the most familiar and popular terms to denote the three-fold contents of the notion of the word "Universe".

The author is fully aware of the inconvenience caused by using two terms (the Universe and the Metagalaxy) in one book for one and the same object, i.e., the observable range of space. However, since many current monographs use the term "Universe" (albeit inadequately), it should not be abandoned completely. The duality of the terminology used in this book reflects a transitory situation in which the identity of the observed part of space (Friedmann's Universe) with all being has not yet become archaic dogma.

3.8.2 Metagalaxy Formation

Hence, we arrive at the question of what the Metagalaxy was like at $t_u = 0$ and the Universe, at $t_u < 0$. This statement of the problem is equivalent to asking how metagalaxies are formed and what was happening in our Metagalaxy at the times $t_u \lesssim 10^{-35}$ s.

3.8.2.1 The Notion of Numerous Metagalaxies

In the early 1960s (perhaps in connection with the successes in explaining baryonic asymmetry in terms of the theory of grand unification)[12] the *Weltanschauung* of physicists was marked by a deep psychological change. Whereas some time ago the very inquiring into the origin of the Metagalaxy was considered senseless because no experimental approaches to the evolution of the Universe at small t_u were available, the explanation of baryonic asymmetry immediately initiated a flood of papers concerned with the origin and very early stages of evolution of the Metagalaxy.[13] For the author, though, the turning point in public opinion was signified not by that flood of papers, some of which contained novel and beautiful ideas, but rather by a singular fact that instantaneously highlighted the situation. In the middle of 1982, the author attended a seminar on the early stage of the evolution of the Metagalaxy; the speaker began his talk with the words: "As everybody now knows, there is a multitude of metagalaxies."[14] When the author heard these words, he felt like a stranger at somebody else's feast. Previously, it had seemed to him that the existence of many metagalaxies had

---

[12] Remember that baryonic asymmetry is a consequence of the baryonic number not being conserved (a particular case of such non-conservation is the decay of the proton) as well as of the charge/parity violation (the CP violation), which properties are inherent in certain variants of the grand unification.

[13] The Soviet physicists A.D. Linde and A.A. Starobinskii as well as A. Guth (USA) and S. Hawking (Great Britain) significantly contributed to the discussion of this problem.

[14] Henceforth, we shall use the word "metagalaxy" to denote objects similar to our Galaxy, but not identical with it.

been his private secret, but it turned out to be an open secret! The author had guessed, however, that the notion of the Metagalaxy originating from a point which arose from "nothing" was an absurdity; something – a background – must have been there prior to the birth of the Metagalaxy. But if there had been a background, then it is only natural to assume that the formation of the Metagalaxy was not a unique event. However, thoughts of this kind were not fashionable at the time, so the author had chosen not to publicize them. And, certainly, it was not reasoning of this nature, but rather serious progress in the theory of the evolution of the Metagalaxy toward the description of the singularity ($t_u = 0$), which caused the change in the physicist's picture of the world; this change stimulated the construction of models of the formation of metagalaxies.

3.8.2.2 Modelling Metagalaxies

The first question to arise is how the coexistence of many metagalaxies can be conceived. The answer can be most easily modelled by imagining an infinite three-dimensional Euclidean space including a system of spheres of various diameters, partly intersecting and partly unconnected topologically (isolated spheres). Each sphere "corresponds" to a metagalaxy. Of course, although such a picture is quite illustrative, it is undoubtedly a serious oversimplification. Metagalaxies may be nonspherical; they may have various dimensions or differ in some other characteristics.

Let us further consider modelling the formation of metagalaxies. Imagine a space filled with a physical vacuum, i.e., with a specific medium homogeneous on the average. In this medium, fluctuations occur, similar to a ripple on the surface of still waters. This "ripple," once isolated from the background, obeys its own intrinsic laws and evolves into something similar to our Metagalaxy. The generation of closed metagalaxies can be most readily modelled; however, according to the theory of gravitation, the total energy of a closed metagalaxy is zero.[15] Therefore, the generation of such a metagalaxy does not require any energy influx from the outside: a closed metagalaxy can arise spontaneously out of a vacuum ("a ripple arises on the surface of still waters at the slightest breeze"). Thus, currently (1983) the opinion prevails that there are an (infinitely) large number of metagalaxies.

But if there are many metagalaxies, one cannot help asking why, in our Metagalaxy, the set of fundamental constants, necessary for the formation of its complex structure, holds. One apparent explanation goes back to Dirac who long ago, in 1937, proposed a hypothesis on the variation of the fundamental constants during the evolution of our Metagalaxy. Of course,

---

[15] The mass of the Metagalaxy is compensated for by the gravitational attraction energy.

at that time, Dirac did not have access to the extensive information available now; he was primarily guided by a fact that struck his imagination, namely, by the smallness of the ratio $\alpha_g/\alpha_e \sim 10^{-36}$. Dirac believed that all nondimensional fundamental quantities must be ~ 1, and that the smallness of the ratio α_g/α_e is characteristic of only a limited range of our space-time.

The originality of the idea as well as Dirac's authority initiated numerous experimental investigations concerned with possible temporal variation of the fundamental constants. The most precise results have been obtained by investigating the rich uranium deposits in Oklo (Gabon, Africa); these results have demonstrated no tendency of the constants to vary with time. We refer to some results stemming mainly from the book entitled *Radioactivity and Evolution of the Universe* by V. Chechev and Ya. Kramarovsky (Nauka Publishing House, Moscow, 1978) which yield;

$$|\dot{\alpha}_e/\alpha_e| < 10^{-17} \text{yr}^{-1} \quad ,$$
$$|\dot{\alpha}_s/\alpha_s| < 5 \times 10^{-19} \text{yr}^{-1} \quad ,$$
$$|\dot{\alpha}_w/\alpha_w| < 10^{-12} \text{yr}^{-1} \quad ,$$
$$|(h\dot{c})/(hc)| < 10^{-12} \text{yr}^{-1} \quad ,$$

and

$$|\dot{\alpha}_g/\alpha_g| < 10^{-10} \text{yr}^{-1}.$$

The dot denotes differentiation with respect to time. The age of the Universe being $\sim 2 \times 10^{10}$ years, these figures mean that within the current cycle of the evolution of the Universe, e.g., the constant α_e changed not more than by 10^{-7} and the constant α_s, not more than by 10^{-8}.

Hence, the experimental data refute Dirac's idea, but with one restriction. The main bulk of the data pertains to the epoch relatively close to the present one. According to geological data, the uranium deposits are some 10^9 years old; this time is short compared to $\sim 2 \times 10^{10}$ years, the age of the Metagalaxy. For a billion years the constants remained unchanged, but what happened before that?

Only few direct data on the variation of the constants are available. There are indirect indications of the invariability of the constants down to small values of t_u, however. The calculations of the cosmological helium abundance (Sect. 2.7) were carried out using the current values of the constants; the results are in good agreement with experiment, thus indicating constancy of the constants back to $t_u \sim 0.1$ s.

More problematic is the argument based on the quantum field theoretical interpretation of the baryonic asymmetry of the Metagalaxy (cf. Sect. 2.6), for it contains a hypothetical element itself, namely, the reality of the theory of grand unification. Nevertheless, the extreme beauty and the uniqueness of this idea of explaining baryonic asymmetry suggest that it should be considered very seriously. Then one could conclude that the constants remained unchanged, beginning from as early a time as $t_u \sim 10^{-35}$ s. It is absolutely unclear, however, what was happening at $t_u < 10^{-35}$ s, and especially at $t_u \lesssim 10^{-43}$ s, when the evolution of the Metagalaxy was determined, according to general belief, by quantum gravitation, for which there is no theory.

3.8.2.3 Structure of the Metagalaxy: The Fundamental Constants

Having made these remarks, we should (and have to) pass on to the principal question, i.e., the reasons for the instability of the structure of the Metagalaxy with respect to the numerical values of the fundamental constants.

It seems that only one answer to this question can be given. Two considerations give a hint: the fluctuational nature of the constants for our Metagalaxy (cf. Sect. 3.7) and the possible multiplicity of metagalaxies. There are no indications that the initial conditions for the formation of metagalaxies are identical. On the contrary, our entire experience, gained by physical experiment (though on a different scale), suggests that disturbances always have some distribution. One may assume, for example, that the "ripple" amplitude is distributed according to some currently unknown law. It is reasonable to believe, then, that a set of the fundamental constants formed in the nascent metagalaxies will not be the same for all of them. A natural modification of this hypothesis is the assumption that our Metagalaxy goes through many cycles. With the beginning of each cycle, a different set of the fundamental constants applies.

There are several aspects of the distribution of the inital conditions for metagalaxy formation:

i)   the *geometrical aspect* concerns the space in which the Universe evolves; the basic characteristic of this space is its dimension;
ii)   the *dynamical aspect* is associated with the interaction laws; mathematically it involves postulating the transformation group for the dynamic equations;
iii) the *algebraic aspect* involves the numerical values of the fundamental constants.

Although all three aspects appear as independent elements in traditional physics, there are grounds for considering them interdependent. A striking example of such mutual dependence is Einstein's theory of general relativity in which the geometrical and the dynamical descriptions of bodies subjected to gravitation are equivalent. This theory is the best-known example of the realization of a program combining two elements of physics — geometry and dynamics.

There are now good prospects for materializing a more general program, viz. the geometrization of the whole of dynamics. This would require introducing relatively complex geometric constructions — stratified spaces. By introducing more sophisticated constructions, one will probably be able to describe all of dynamics in geometrical terms.

Thus, real progress in combining dynamics and geometry is now in sight. The third basic element of physics, the fundamental constants, should not remain unaffected by these unification tendencies. Concrete advancements in unifying all three of these elements of physics have admittedly been much more modest. Although papers have appeared, attempting to derive the numerical values of certain fundamental constants from geometrical considerations, by and large, they seem unsatisfactory. The most general version of a unification program would be as follows.

Some physical space — the vacuum of a large (infinite?) dimension — exists. Spontaneous fluctuations of this vacuum generate universes having their own dimensions, sets of interactions (i.e., groups of transformations of the dynamic equations), and numerical values of the fundamental constants. All these elements of physics are interrelated. The geometry and the dynamics are different forms of the description of the evolution of physical systems. The spaces, which are an arena of particle interactions, may generate numbers — fundamental nondimensional constants — themselves; the simplest example of such a number is the dimension N. Another example borrowed from topology is connectivity, i.e., the number of regions, any two points of which can be connected by a continuous line. A simply connected region is exemplified by a three-dimensional sphere; an example of a double connected space are two nonintersecting spheres, etc.

The above examples are most elementary. However, in geometry there are numerous relationships between the characteristics of spaces and numbers. It appears that the choice of a physically meaningful relationship between geometry and the fundamental numbers is one of the principal problems in physics. Solving this problem may lead to a unification of all three elements: geometry, dynamics, and the fundamental constants.

Certain fragments of the program to include the fundamental constants in the dynamics of the expanding Metagalaxy have been realized. The Soviet physicists D.A. Kirshnits and A.D. Linde have propounded a particular scheme in which the unified interaction constant α_u is split up into the interaction constants of the individual interactions. At the temperature $kT_u \sim m_{wesg}c^2 = 10^{19}$ GeV($t_u \sim 10^{-43}$ s), the gravitational interaction splits off; at $kT_u \sim m_{wes}c^2 \sim 10^{15}$ GeV($t_u \sim 10^{-35}$ s), the strong interaction splits off; at $kT_u \sim m_{we}c^2 \sim 10^2$ GeV($t_u \sim 10^{-10}$ s), the electroweak interaction splits off. All the interactions split off individually (cf. Chap. 4).

Although no complete theory of the formation of the mass of the elementary particles is available as yet, the majority of physicists are convinced that the masses and the α-constants are related to each other. If this is true, and if the Kirshnits-Linde scheme does describe reality, then of all the numbers governing the structure of the Universe which are given on p. 88, only two remain independent: the dimension N and the unified interaction constant α_u. These two numbers have to be generated by geometry (by space).

Now it is time to ask a final question: What is the significance of the instability of the structure of the Metagalaxy with respect to the values of the fundamental constants? First of all, the very existence of the Metagalaxy with a unique set of fundamental constants points to the uniqueness of our Metagalaxy; otherwise it is difficult to explain the occurrence of a singular set of constants necessary for the rise of complex forms of the matter. A simpler version of this interpretation would be the assumption of a number of Metagalaxy cycles.

It should be emphasized that the analysis of the existing distributions suggests that the set of fundamental constants occurring in our Metagalaxy is a very strong fluctuation. This conclusion is substantiated by a comparison of the fundamental constants with other constants demonstrating: (i) the extreme smallness of the value of the mass m_e as compared with $\langle m \rangle$; (ii) the smallness of the value of Δm_N as compared with $\langle \Delta m \rangle$; (iii) the smallness of the value of ε_d as compared with $\langle \varepsilon \rangle$; (iv) the smallness of the ratio m_p/m_{wes} as compared with unity; and (v) the smallness of the ratio α_g/α_e compared with unity. This notion of the "formation" of the fundamental constants is also supported by the following fact: a number of relations between fundamental constants have been presented above (cf. (3.30–32)); these relations are necessary for the formation of galaxies in a closed Metagalaxy and for the rise of bright stars (condition (3.36)). All these conditions are fulfilled, but with no margin. Possibly, this implies that we are living in a closed Metagalaxy. The criticality of these relations (i.e. their being fulfilled just in the limit) then also indicates the fluctuational character of the formation of the constants. This circumstance has to be accounted for by any theory claiming the interpretation of the numerical values of the fundamental constants.

Our Metagalaxy is probably a giant fluctuation among other universes (as far as the numerical values of the fundamental constants are concerned). This fluctuation is the basis of the complex structure of the Metagalaxy.

4. The Beginning and End of the Metagalaxy

4.1 Updating our Knowledge of the Metagalaxy

Since the completion of the foregoing chapters (1983–1984), the problem of the formation of the Metagalaxy has come to the fore of modern physics. This fourth chapter gives an account of the progress made since 1984. The conclusion reached in Chap. 3 (i.e., that many metagalaxies coexist) will be shown to have been confirmed.

Inevitably, the reader will be referred to material contained in Chaps. 1–3, since Chap. 4 is concerned with interrelated subjects touched upon there: the notion of a physical vacuum; the complex nature of physical space; and the origin of the numerical values of the fundamental constants.

4.2 Describing the Metagalaxy

As in Chap. 3, we shall refer to the Metagalaxy as the observable part of the Universe, and the metagalaxy as the hypothetical objects beyond the Metagalaxy. Study of the Metagalaxy involves both the familiar forms of matter (substance, radiation) and those less familiar (the physical vacuum).

To proceed with a discussion of particular problems of modern cosmology, it is necessary to decide upon the terminology to be used. This is not easy, and often, unfortunate terminology leads to misunderstandings, sometimes even in solid monographs.

Since this essay deals with the problems of the origin of the observable world, to avoid confusion, it is necessary to explain what is meant by "observable world." The entire existing material world, or Universe, is boundless in time and space, and it is infinitely diverse with regard to the forms of matter. However, we are primarily interested in the observable material world including both the well-studied forms of matter (such as particles and radiation) and those which only recently became the object of study (e.g., the physical vacuum). This observable part of the Universe is usually referred to as the Metagalaxy.

The Metagalaxy is made up of various structural elements (stars, galaxies, galaxy clusters, and superclusters, etc.). Like any empirical object, it is characterized by certain parameters; in particular, it has a definite size (radius) which is of the order of 10^{23} km.

It should be emphasized that this size of the Metagalaxy is tremendous as compared with the scale directly perceptible. Man has travelled the length and breadth of the Earth (whose size is of the order of 10^4 km) and has visited the Moon, 4×10^5 km away. The first unmanned cosmic apparatuses have crossed the orbits of the most remote planets of the solar system and are currently some 10^9 km away. This last value is related to the size of the Metagalaxy as a distance of 1 cm relates to this very value of 10^9 km, however.

In constructing models of the Metagalaxy, one has to make extrapolations beyond the Metagalaxy proper, in that the structure of the Metagalaxy as well as the known laws are extended beyond its boundaries.

4.3 The Universality of the Physical Laws

An important question when approaching the subject of the Metagalaxy is whether it is legitimate to extrapolate the laws studied on Earth, on such a small scale, to the Metagalaxy as a whole. Newton's law of gravitation, which claimed to be a universal law, was restricted, at Newton's time, to the solar system. However, since then, it has been substantiated for multiple (double) systems as well as for clusters of cosmic objects, such as stars and galaxies, and one can hardly doubt its truly universal applicability.

Similarity between luminosity spectra of stars, studied in great detail, also supports the universal character of the electromagnetic interaction in the entire Metagalaxy. It is much more problematical, however, to experimentally verify the universality of microscopic interactions (the strong and weak ones; cf. Sect. 1.4) beyond the Earth. The first attempts to observe neutrinos arising in the Sun have already demonstrated that there is something in the structure of our star or in the strong or weak interaction that we do not understand. It has turned out that the observed flux of the solar neutrinos is much smaller than that calculated with the standard model of the Sun, using the properties of the strong and the weak interactions obtained in laboratory tests.

This discrepancy must now be explained by the physicists who discover these laws in the laboratory. Rather than revising the principles well studied on Earth, it will be necessary to discuss new hypotheses which do not contradict the laboratory measurements. Initially, the attempts to solve the "neutrino crisis" were based on modifying our view of the structure of the

Sun. Later on, an alternative hypothesis was propounded by invoking the existence of an extremely small neutrino mass (about $1\,\text{eV} \simeq 10^{-36}\,\text{kg}$). The existence of such a neutrino mass would not affect any laboratory results, but would lead to a decrease in the neutrino flux on the way from the Sun to the Earth, thus explaining the neutrino deficiency.

Why, however, did physicists not choose a more radical, but simpler, way by assuming, for instance that the weak interaction responsible for particle decay and the rise of neutrinos is somewhat weaker in the Sun than on the Earth? There is certainly no unequivocal answer to this question. However, as mentioned above, physicists presume that the fundamental physical laws are universal, i.e., their validity within the Metagalaxy does not depend on position or time. Furthermore, a number of experimental arguments weigh heavily in favor of this viewpoint.

Thus, there is experimental evidence that the gravitational and the electromagnetic interactions are practically universal. It appears natural to extend this property to microscopic interactions as well. On the other hand, observations (and not just speculative arguments) indicate that the space of the Metagalaxy is nearly isotropic and uniform, the Earth as a physical object not being a singular point within it (a principle first propounded by Copernicus). It is therefore quite natural to expect that the physical laws will be the same on the Earth and beyond it, i.e., within the Metagalaxy.

It is this kind of reasoning which underlies the assumption of the universality of the physical laws. Still, in spite of the persuasiveness of these arguments, ideas emerge in astronomy which fundamentally change established physical conceptions, like the "creation" of matter out of nothing in the Metagalaxy or the existence of a significant fraction of antimatter. However, up to now such "novelties" have been short-lived: they have been discarded on the basis of experimental evidence.

Thus, it follows that the physical laws are probably universal in time and space within the Metagalaxy. This is true with one proviso, however. A perfect experimental substantiation only refers to the time scale of 1 billion years reckoned from the present epoch, while the lifetime of the Metagalaxy is about 10 to 20 billion years.

4.4 The Very Beginning

There are indirect data indicating the invariability of the physical laws during almost the entire evolution of the Metagalaxy, or, more precisely, back to the time $t \approx 1\,\text{s}$ from the beginning of the Metagalaxy. But what was happening at the time $t \ll 1\,\text{s}$ or — a more daring question — at $t = 0$?

Here, the experimental evidence is scarce and indirect. However, just touching on the secrets of the emergence of the Metagalaxy, i.e., analyzing the physical situation at $t = 0$, is fascinating, for it offers the possibility of lifting the veil of mystery covering the birth of the Metagalaxy (which is sometimes inadvertently identified with the Universe).

At first sight, even to attempt such an analysis seems pointless. Indeed, the Metagalaxy is a unique object, and one is not in a position to reproduce the process of its creation in a laboratory in order to test theoretical models. This is true with one reservation. In experiment, quite a few distributions of various constants (e.g., of the masses of elementary particles) have been investigated in great detail. Some of these constants play the central, fundamental role in the structure of the Metagalaxy, while others (the majority) have practically no effect on its evolution. It was found that these constants usually have very pronounced values within the experimental distributions. This fact can be used as a basis for an analysis of the evolution of the Metagalaxy soon after its "birth" at $t = 0$.

During the complex process of acquiring knowledge about the origin of the Metagalaxy, one cannot, of course, avoid making assumptions reaching beyond the limits of experimental fact. However, we shall try by all means to clearly distinguish between firmly established experimental fact and subsidiary hypotheses.

4.5 Models of the Metagalaxy

Over the years, investigation has demonstrated that the Metagalaxy can be characterized by two fundamental properties (often referred to as the Cosmological Principle): The Metagalaxy is isotropic and uniform (cf. Sect. 2.2).

One reservation is due at this point. The Cosmological Principle (cf. Sect. 2.2) was introduced by theoreticians as a certain idealization to permit simplifications of theoretical modelling. Obviously, this principle is violated at a scale which is small compared with the size of the Metagalaxy; after all, stars, galaxies, clusters of both, and finally a large-scale structure of the Metagalaxy exist. Still, in analyzing the evolution of the Metagalaxy as a whole on the basis of this principle, these violations are usually neglected. There is an important empirical justification for doing so. In fact, the isotropy of the Metagalaxy has been demonstrated by numerous observations of cosmic background radiation (cf. Sect. 2.5). Investigation of its temperature T in all possible directions has shown its constancy within the accuracy of modern measuring devices. This accuracy is quite high: $\Delta T / T \sim 10^{-1} - 10^{-4}$, ΔT being the experimental error.

Hence, high-precision measurements have not revealed any deviations from isotropy. The uniformity (homogeneity) of the Metagalaxy is more intricate. It is obvious that on a small scale, the Metagalaxy is inhomogeneous (cf. the very existence of the solar system). The existence of such inhomogeneities as galaxy clusters measuring 10^{19} km is now well established; there are probably also superclusters whose characteristic size is 10^{20}–10^{21} km. The latter quantity is the largest size of the inhomogeneities detected. It is therefore natural to assume that the Metagalaxy is homogeneous with the accuracy given by the ratio of this quantity to the radius of the observable part of the Universe, i.e., with the accuracy of 10^{-3}–10^{-2}.

One may have various opinions about the significance of this figure. As a rule, one tends to think that the superclusters do not seriously disturb the homogeneity of the Metagalaxy, or, more precisely, do not influence its evolution. Accordingly, they can be neglected when analyzing the development of the Metagalaxy as a whole.

In a certain sense, a well-known joke about the Gaussian statistical law can be paraphrased to refer to the homogeneity of the Metagalaxy: "The mathematicians consider it an empirical fact while the physicists deem it a mathematical theorem." Be that as it may, the Cosmological Principle makes it possible to draw quite general conclusions about the evolution of the Metagalaxy without invoking complex physical arguments based on the theory of relativity.

The so-called Hubble law, stating that the relative velocity, v, of two cosmic objects is proportional to the distance between them, follows from the Cosmological Principle (cf. Sect. 2.2). H is independent of the distance between the objects but changes with time. (Remember that the time t is reckoned from the beginning of the Metagalaxy.) Hubble's law can be written in the form $v = H(t)r$.

The last equation that connects the characteristics of the Metagalaxy has several major classes of solutions. In order to keep our presentation free of tedious mathematical detail, we only give the final results for the major classes of solutions:

$$H = 0; \quad r = \text{const} \quad ; \tag{4.1}$$

$$H = \text{const} \neq 0, \quad r = r_0 e^{Ht} (r_0 = \text{const}) \quad ; \tag{4.2}$$

$$H = a/t, \quad r = bt^a (a, b = \text{const}) \quad . \tag{4.3}$$

In the above solutions, the quantity r can be given two different interpretations. It can be considered as the relative distance between two arbitrary points (as a scale factor). On the other hand, a more simplified, and thus not quite precise, interpretation of r would be by identifying it with the Metagalaxy radius, $r = R_M$.

The solution represented by (4.1) corresponds to Einstein's model of the Metagalaxy. The physical meaning of this solution is quite obvious: the Metagalaxy is static, all its parameters (the distance r between two material "points", e.g., galaxy clusters, the average density of matter ϱ, and other characteristics) not varying with time.

The de Sitter model corresponding to solution (4.2) is no longer static. The distance between two arbitrary objects changes very rapidly (exponentially); the density ϱ, however, can be shown to remain constant ("stationary Metagalaxy"). At first glance, it appears that the two properties of de Sitter's model are contradictory. We are accustomed to thinking that changing the volume of a system is associated with a change in its density. For this reason, the de Sitter model has not been used for a long time. Only recently has it again drawn attention, but under a new premise. (We shall come back to this issue below.)

Finally, solution (4.3) corresponds to the nonstationary model proposed by Friedmann (cf. Sect. 2.2).[1] Within that model, both the scale factor r and the density ϱ vary with time.

We now confront one of the principal questions in cosmology: Which of the models of the Metagalaxy is an adequate representation of reality? There is no a priori answer to this question; it is up to experiment to give an answer.

Experiments starting with Hubble's numerous observations from 1929 on have demonstrated that our Metagalaxy is nonstationary. The distance between the galaxies increases. A proof of this statement has been obtained by measuring the radiation spectra of the galaxies. The spectra turned out to be shifted toward the red side (galactic redshift), indicating recession of the galaxies, i.e., increase of the scale factor. Hence, the model proposed by Einstein contradicts the experiment.

As already mentioned, de Sitter's model is also in contradiction to common knowledge. Furthermore, it can be shown that this model leads to a very unusual equation of state, $p = -\varepsilon$, where p is the pressure and ε is the energy density of matter. This equation of state is not fulfilled for any form of matter studied in the laboratory; neither is it fulfilled for radiation obeying the equation of state of the form $p = \varepsilon/3$. Consequently, de Sitter's model is not consistent with reality, at least in the current epoch of the

---

[1] Generally, Friedmann's solution takes a more complex form which reduces, however, to (4.3), in particular, in the case of small t, of primary interest here. It should be emphasized that we have omitted complicated calculations within the general theory of relativity, although the approaches of Einstein, de Sitter, and Friedmann were based on it. The postulates of the Cosmological Principle are so strong that the possible character of the evolution of the Metagalaxy is predetermined by them.

evolution of the Metagalaxy. One is led to conclude that the only model capable of describing reality is Friedmann's; a more detailed account of it is given below.

4.6 The Friedmann Model (cf. Sect. 2.2)

So much has been written about the Friedmann model that it might appear superfluous to dwell on it. However, general acceptance of this model has obscured some of its difficulties. Resolving them would be beneficial with regard to the problem of interest here, namely, what was happening in the early Metagalaxy at $t \approx 0$. That is why our discussion of this model will concern not only its indisputable merits, but also its difficulties. (The latter will be discussed in a subsequent section.)

4.6.1 The Long Way to Recognition

It would be instructive to give a brief account of the dramatic and tortuous path the Friedmann model took toward recognition by the scientific community. In 1922, right after the publication of Friedmann's first paper on cosmology, Einstein wrote of it: "The results (obtained by Friedmann — *author's remark*) on the nonstationary world appear to me suspicious." But one year later, Einstein, a man of principle, wrote something different: "I consider the results of Mr. Friedmann correct and throwing a new light." In the subsequent editions of his book "The Meaning of Relativity", Einstein always based his presentation of cosmology on Friedmann's ideas.

But the pendulum of public opinion had only been set in motion by Einstein. It was not until Georges Lemaître confirmed Friedmann's main result on nonstationarity, that his ideas won general recognition. (In his cosmological model, Lemaître considered the more complex case of a physical medium.) Ultimate recognition was obtained in 1929 when Hubble discovered the redshift of galaxies (cf. Sect. 2.1).

Friedmann's ideas of a nonstationary Metagalaxy unexpectedly became involved in vulgar social doctrines; experimental errors[2] and some imprecise formulations in his fundamental works contributed in leading some scientists to declare Friedmann's theory "pseudo-science".[3]

---

[2]   Initially, Hubble's constant was determined with a very large error leading to an underestimation of the lifetime of the Metagalaxy. This was in contradiction to the lifetime of the stars as determined experimentally: the latter was one order of magnitude larger than the Metagalaxy lifetime according to Friedmann. This was, of course, a great blow to his model.

[3]   Fuel to the fire was added by an incidental circumstance: Abbé G. Lemaître (who later became President of the Academy of the Vatican) placed himself at the head of a group of supporters of the theory.

This strange situation did not change until the early 1950s, when the magnitude of Hubble's constant was determined with much more precision. This allowed establishing an order-of-magnitude agreement between the lifetime of the Metagalaxy and that of the stars. Thereafter, Friedmann's ideas won general recognition; certain cosmologists even declared them the "ultimate wisdom". Further development of cosmology now appeared only conceivable in the framework of Friedmann's theory. Only in the late 1970s was a more realistic attitude toward Friedmann's cosmology adopted and some attempts at revision of certain statements of his theory were made.

The "canonization" of the Friedmann model was promoted by significant observational confirmation. First, cosmic background radiation, predicted on the basis of the Friedmann model, was discovered (1965). And second, progress in nuclear physics has made the confirmation of the abundance of helium in the Metagalaxy (in accord with Friedmann's cosmology) possible. This quantity (about 25 % by mass) is in excellent agreement with the observational data. It should be noted here that the theory of helium synthesis refers to very short times in the development of the Metagalaxy, $t \approx 0.1\,\mathrm{s}$ to $5\,\mathrm{min}$. In addition, the calculation of the helium abundance was based, at that time, on the assumption that the laws of nuclear physics hold down to such times.

Even more impressive was the explanation of baryonic asymmetry proposed in the Soviet Union in the late 1960s by A.D. Sakharov. It should be remembered that the notion of baryonic asymmetry of the Metagalaxy refers to the presence of protons and electrons, while antiprotons and positrons are practically absent (cf. Sect. 2.6). These ideas came to bear in the grand unification theories.

Within these theories, a superheavy particle, the X-boson exists, whose mass is of the order of $10^{15}\,m_\mathrm{p}$ where $m_\mathrm{p} \approx 10^{-24}\,\mathrm{g}$ is the proton mass (cf. Sect. 2.6).[4] This particle should have a striking property: it must decay into a proton and an electron with a probability somewhat higher than that of the antiboson $(\overline{X})$ decay into an antiproton and a positron. It is this small bias that gives rise to the occurrence of baryonic asymmetry. All these processes involving the X-boson decay take place at very high temperatures of the order of 10^{28} K. It should be emphasized furthermore that the time at which the X-bosons are formed and the baryonic asymmetry arises turns out to be inconceivably small: of the order of 10^{-35} s.

From the grand unification theory, one can derive an order-of-magnitude estimate for the number that characterizes the baryonic asymmetry of the Metagalaxy, namely, the ratio n_p/n_γ of the average concentrations of primeval protons and photons in the Metagalaxy. The experimental value of this

---

[4]   It should be noted that owing to its very large mass, the X-boson cannot be detected by accelerators.

ratio ranges from 10^{-8} to 10^{-10} and the theoretical one, from 10^{-5} to 10^{-10}.

One achievement of Friedmann's cosmology should be mentioned once again. In Friedmann's model, the lifetime of the Metagalaxy is $t \approx 1/H_0$, where H_0 is Hubble's constant corresponding to the current epoch. By using the observational data on the constant H_0 it can be found that $t \approx 10\text{--}20$ billion years. This figure is in agreement with the age of the Earth (4.5 billion years) and of old stars (about 10 billion years) measured directly, on the basis of radioactive decay.

4.6.2 Difficulties

The success of Friedmann's cosmology had a magnetizing influence on the research dealing with this area of science. Certain principal difficulties of the theory did not directly contradict the observational data; thus, they were either ignored or dealt with in a speculative way, within Friedmann's theory itself. Here we shall dwell on two difficulties of Friedmann's model.

4.6.2.1 Singularity

From solution (4.3) of the form $R_M = bt^a$, it follows that a finite energy is concentrated within zero volume. Hence, at $t = 0$, the energy density ε becomes infinite. A state in which a certain physical parameter goes to infinity is referred to as a *singularity*.

The singularity situation in cosmology is in contradiction to all accumulated physical experience. Under terrestrial conditions, a phase transformation of matter always occurs as $\varepsilon \to \infty$, so that the singular state is not achieved. To reconcile this fact with the existence of the singularity, one attempted to change the initial conditions of the Friedmann model. One version was to abandon, at $t = 0$, the Cosmological Principle by admitting the anisotropy of the Metagalaxy at the beginning. However, even such strong auxiliary postulates were not able to solve the problem of the singularity. It was shown by British astrophysicists that the singularity persists, provided that the condition $\varepsilon + p > 0$ is fulfilled.

This condition is quite natural and it is fulfilled for all forms of matter well studied in the laboratory. It reflects the simple fact that on increasing the pressure p, the compression (and hence the energy density ε) also increases. Otherwise, matter would be unstable and would start decaying into different phases, and the *energy dominance condition* would be restored. The Singularity is thus inherent in the Friedmann model.

4.6.2.2 The Horizon (cf. Sect. 2.8.3 and Fig. 2.5)

According to the theory of relativity, no information can be transmitted with the speed ot light, c. Consequently, at a time t elapsed after a signal has been emitted, only regions within a distance $r < ct$ can be causally connected. When applied to the Metagalaxy, this statement means that a signal emitted even at the very moment of its formation can only reach parts within $r \leq ct$ (t being the age of the Metagalaxy). Hence, two arbitrary regions can be causally interrelated only if they satisfy this condition.

The distance ct is the limit of information transmission between two "points," in particular, between the solar system and a galaxy. The distance ct is referred to as the horizon. Generally speaking, the dimensions of the horizon and of the Metagalaxy need not be identical. The horizon places a principal limitation on the observability of cosmic objects, and no sophistication in instrumentation can make it possible to visualize objects from beyond the horizon.

The size of the horizon is a quantity that has dual significance, for it is determined by the age of the Metagalaxy, t. In Friedmann's cosmology, t is proportional to $1/H$; on the other hand, this time is equal (within an order of magnitude) to the age of old stars. Currently the horizon of events is about $ct \approx 10^{23}$ km, which coincides with the observable size of the Metagalaxy. However, this coincidence only applies to the current time.[5]

To clarify the reasoning below, let us consider the following argument. Assume the Metagalaxy expands slowly enough for this expansion to be neglected for a while. Now consider the situation at a time $t' < t$. At the moment t', the size of the Metagalaxy can be taken to be unchanged, while the horizon is smaller than the size of the Metagalaxy, owing to $ct' < ct$.

According to solution (4.3), the size of the Metagalaxy varies as a power function, $r = t^a$, with $a < 1$ (cf. Sect. 2.2, 2.3). Therefore the size of the Metagalaxy changes more slowly than the size of the horizon, which is proportional to t. This means that if the sizes of the Metagalaxy and of the horizon currently coincide, the size of the horizon was once smaller: at $t' < t$. Consequently, this model implies that the early Metagalaxy was subdivided into a number of regions not causally connected.

At first glance, this fact appears to be quite innocent. It turns out to be a dangerous problem, however, if considered together with the extreme isotropy of cosmic background radiation. How was it possible that one part

---

[5]   The value of 10^{23} km for the size of the horizon follows from the relation $t = 1/H$, characteristic for solution (4.3) which, as already mentioned, is always valid for small t. In considering the current epoch of the expansion of the Metagalaxy, the form of this relation may be different, yielding a value of the horizon which may be larger or smaller than 10^{23} km. This is, however, not very important for the considerations below.

of the Metagalaxy, not causally related to another one, adjusted to it to jointly produce an isotropic (spherical) geometry?

The problem can be considered in a somewhat different way. Take any two directions separated by a sufficiently large angle. The galaxies observed along these directions originate from two different regions not interrelated in a distant past. Why, then, is an equal number of galaxies observed along the two directions?

The rest of our presentation will largely deal with describing the attempts at solving the above-mentioned difficulties of Friedmann's model, in particular, the problem of the horizon. But first we shall have to digress from the immediate cosmological aspects.

4.7 The Physical Vacuum

Unfortunately, there are two notions associated with the term "physical vacuum." The first one, familiar from high school texts and best known, is related to rarefied gases. The second one, adopted by physics relatively recently, is an organic part of quantum field theory. Physical vacuum in the latter sense means that a system does not contain *real particles* of a given kind (e.g., positrons).

In quantum field theory, the notion of a physical vacuum is broader, however. According to this theory, *virtual particles* also exist whose "lifetime" Δt, determined by the uncertainty principle, is extremely small. For instance, for particles with the electron mass m_e, this time is $h/m_e c^2$ (where h is the Planck constant). Inserting the numerical values, one obtains $\Delta t \approx 10^{-21}$ s.

One might think that introducing a set of virtual particles does not contribute much to the clarification of the nature of the physical vacuum; indeed, even the virtual particles would not arise in the absence of a particle reservoir. And the hypothesis of the existence of such a reservoir leads to a further question: Why don't the virtual particles transform into real, long-lived particles?

In trying to answer these questions, it would be useful to review the development of elementary particle physics, in particular, by recalling Dirac's famous paper predicting the existence of the positron (1928). In that paper, the existence of an infinite numer of *nonobservable* electrons with negative energy was postulated. The set of these electrons was named the *electron (positron) vacuum*. A positron can be viewed as a "hole" in the vacuum in which all admissible states are filed. To produce such a "hole" (i.e., a positron) energy exceeding $2 m_e c^2 \approx 10^{-6}$ erg must be provided. This energy is large on the scale of microphysics, and consequently the virtual particles do not spontaneously convert to real ones.[6]

At first glance, this picture postulating the existence of an infinite number of nonobservable entities might appear the product of a morbid imagination. However, this phantasy was a logical consequence of quantum mechanics and the theory of relativity. And this logic proved triumphant: in 1932, the positron was discovered, in strict agreement with Dirac's theory, although between 1928 and 1932 most of the physicists had considered Dirac's prediction a delusion. Later on, the necessity of taking the Dirac's vacuum into account when calculating elementary particle interactions was demonstrated.

A particular substantiation of the concept of a physical vacuum, almost unique in its precision, was obtained in calculating the energy level shift of orbital electrons due to their interaction with a physical vacuum. As a result of this interaction, a spectral line occurred with the predicted frequency shift $\nu_{\text{theor.}} = 1057.91 \pm 0.01\,\text{MHz}$, while the experimentally measured value of the frequency was $\nu_{\text{exp}} = 1057.90 \pm 0.06\,\text{MHz}$.

Not less impressive was the calculation of the anomalous magnetic moment (i.e., the deviation of the magnetic moment from the so-called Bohr magneton) of the electron, which is in agreement with the experimental value, accurate to eleven (!) digits. The anomalous magnetic moment also stems from the interaction of an electron with a physical vacuum.

Thus, the notion of the Dirac electron-positron vacuum is related to physical reality, as manifested in experiment with incredible accuracy. However, some points remain unclear: for example, the energy density of the Dirac vacuum is infinite. Besides a very awkward divergence contradicting all the traditions in physics, there is an experimental question of great importance involved. It follows from cosmological considerations concerning the speed of the expansion of the Metagalaxy that the average total energy density of matter (including a physical vacuum) can be estimated to be $\leq 10^{-8}\,\text{erg/cm}^3$. This figure is extremely small (cf. the corresponding energy density of water, $\varrho c^2 = 10^{21}\,\text{erg/cm}^3$), and is far from suggesting an infinite value.

It should be noted – not without regret – that this difficulty has not been completely resolved up to now. Theoretically, the problem is expected to be solved along the following lines. Besides the electromagnetic interaction responsible for the Dirac (electron-positron) vacuum, other well-studied interactions exist, viz. the strong, the weak, and the gravitational ones (cf. Sect. 3.3). It is commonly assumed that physical vacua of different interac-

---

[6] By way of illustration, the physical vacuum can be compared, in this sense, with a stone hit by a particle. Only for a very large energy of collison will fragments of the broken stone be produced.

tions (possibly including those not revealed up to now) make contributions which enter the energy density with different signs, so that the total magnitude of the energy of a physical vacuum is very small or even zero.

The author has to admit a certain dissatisfaction with this explanation: the hypothetical element in it is too big. As a matter of fact, one question is substituted here for another one: What is the reason for this mutual compensation? Some considerations regarding the last question will be presented below.

At this point we digress to elementary particle physics. According to current theoretical conceptions, besides the mentioned four interactions, a fifth one also exists, the so-called scalar field. This field mediates the interaction between particles with zero spin,[7] commonly referred to as Higg's particles, and is introduced to explain the existence of mass in elementary particles. At present, it is the only explanation modern theory has been able to come up with. This mechanism, proposed by P. Higgs, is determined by the interaction between hypothetical scalar particles (Higgs' particles).

Although the Higgs' particles have not been detected experimentally (presumably because of their very large mass), the majority of physicists believe in their existence. Therefore their properties have been studied theoretically in great detail. In particular, the characteristics of the physical vacuum of the scalar particles have been investigated.[8]

Thus, by definition, a physical vacuum is a state without real particles but, at the same time, it is a system in which the energy depending on a given field is a minimum. It is known from experience that such a physical vacuum is a very stable entity, while a general stability theory leads to the conclusion that such a state is characterized by a minimum potential energy. In everyday life, one encounters numerous applications of this theorem. Perhaps most illustrative is a downhill sleigh-ride on a trajectory passing through a well. If the height of both sides of the well is large enough, then after several oscillations, the sleigh will land at the bottom of the well, i.e., in the state of minimum potential energy.

Now let us come back to the scalar Higgs field. Theoretical study of this field led to the discovery of a very interesting regularity. At very high temperatures, the dependence of the potential energy U on the scalar field φ has a single minimum corresponding to the only stable state, $\varphi=0$ (cf. Fig. 4.1a). However, at zero field, $\varphi=0$, an interaction with it cannot give

---

[7] Remember that the spin is one of the quantum numbers of the elementary particles. It can take any integer and half-integer values in the units of h (cf. Sect. 1.2.1).

[8] An excellent presentation of the properties of this vacuum is given in the article by D.A. Kirshnits and A.D. Linde entitled "Phase Transformations of Elementary Particles in Cosmology" in: *Science and Mankind,* (Znanie Publishers, Moscow, 1982) p. 165 (in Russian)).

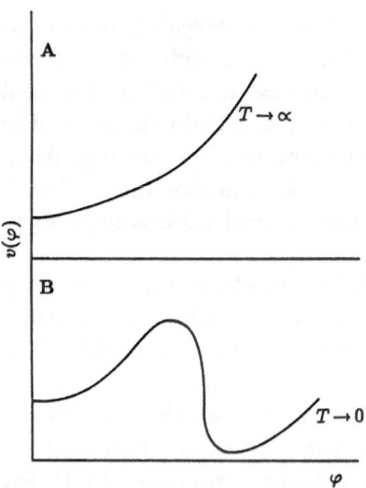

Fig. 4.1a,b. The potential energy U as a function of the scalar field φ for very high (a) and very low (b) temperatures. The minimum at $\varphi \neq 0$ corresponds to the current state of physical vacuum

rise to any massive particles. To produce the mass of particles, a stable vacuum state with nonzero φ is necessary.

Detailed investigations of the temperature dependence of the scalar field potential have demonstrated that at sufficiently low temperature T (as compared to the cosmic rather than the "human" scale, of course), a second minimum appears in the dependence $U(\varphi)$, viz., at $\varphi \neq 0$ (cf. Fig. 4.1b). As theoretical analysis shows, it is only at $\varphi \neq 0$ that the occurrence of elementary particle mass is possible. Certainly, a transition from state 1 (minimum at $\varphi = 0$) into state 2 (minimum at $\varphi \neq 0$) is accompanied by a serious structural change of vacuum (phase transition).

According to the estimates based on theories unifying the different interactions, such a transition should take place at least at two values of the temperature. One of them, 10^{28} K, corresponds to the mass of heavy X-bosons, $m_x \approx 10^{15} m_p$, whose existence is predicted by the grand unification theory, as discussed above. The other transition, involving the splitting of the weak and the electromagnetic interactions, is associated with a temperature approximately equal to 10^{15} K.

4.8 The de Sitter Model: The Beginning of the Metagalaxy

As mentioned before, the scalar particle field has to possess vacuum states as well. These states are associated with systems in which real scalar particles are completely absent. It is not possible to exactly calculate the energy density of the physical vacuum, ε_{vac}, within the theory. One can only give some estimates based on dimensionality considerations. Thus, one may assume that the quantity ε_{vac} has to be determined by the universal constants h, c, and G. Confining oneself to these constants only, one would obtain $\varepsilon_{vac} \approx c^7/G^2 h \approx 10^{115} \text{erg/cm}^3$. This value is tremendous, even with respect to the cosmic scale. Indeed, the estimated energy stored in the entire Metagalaxy is of the order of 10^{80} erg.

One may try to estimate ε_{vac} by using, in addition to the universal constants, the mass of the elementary particles. A typical (average) mass is the proton mass, $m_p \approx 10^{-24}$ g. Together with the constants h and c, this quantitiy readily yields $\varepsilon_{vac} \approx m^4 c^5/h^3 \approx 10^{38} \text{erg/cm}^3$, which is already closer to the cosmic scale but still too large as compared to the observed limit, $\varepsilon_{vac}^{exp} < 10^{-8} \text{erg/cm}^3$.

Let us leave aside the unsolved problem of the smallness of ε_{vac}^{exp}[9] by only saying that the experimental value of the energy of a physical vacuum is appreciably smaller than that expected from rather rough dimensional considerations. The transformation of a physical vacuum at small time t described in the preceding section can be then regarded as a perturbation of the vacuum leading to an abrupt change in the energy density. This transition is illustrated in Fig. 4.2, showing the values of ε_{vac} prior to and after the jump. Jumps in the energy density in phase transitions is a phenomenon quite common in Nature. Typical examples of phase transitions are transformations of water to vapor or ice. Phase transitions are often accompanied by the liberation of a significant amount of energy.

But what happens to the energy released during the phase transition of a physical vacuum? It is quite reasonable to assume that this energy is used in the expansion of the Metagalaxy and the generation of new particles. Special properties of the scalar vacuum permit resolution of the difficulties of the Friedmann model described above.

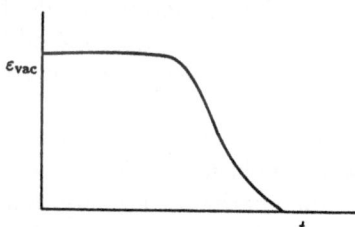

Fig. 4.2. Dependence of the energy density of physical vacuum, ε_{vac}, on the cosmological time t

---

[9] Some considerations explaining the smallness of ε_{vac}^{exp} will be presented below.

One of these special properties can be described as follows: the larger the energy density of a scalar vacuum, the stronger its tendency to accumulate new particles. This property is in contradiction to the common sense; for example, on compressing a gas, its density and pressure increase so that it repels new particles. This property of a physical vacuum is one of the reasons for the nonobservability of vacuum particles in the free state: a physical vacuum does not "release" particles confined in it.

In mathematical terms, this property of a vacuum is expressed in an equation of state quite unusual for laboratory conditions, $p = -\varepsilon$. But it is precisely the equation of state characteristic of de Sitter's model! Consequently, in a physical vacuum, a metagalaxy can and actually must develop, provided that de Sitter's model applies. The assumption of an initial de Sitter stage eliminates the aforementioned difficulties of Friedmann's model. Indeed, within the de Sitter model, the Metagalaxy radius varies exponentially (as e^{Ht}), which eliminates the singularity.

On the other hand, as a consequence of the rapid exponential expansion, the volume of the primary Metagalaxy would exceed the observed volume by many orders of magnitude. (According to some estimates, the Metagalaxy radius in de Sitter's model would be as large as 10^{800} cm). Then the problem of the horizon can be easily solved: in de Sitter's cosmology, the expansion proceeds extremely fast (as compared with the law $r \sim t$). Accordingly, if in the present epoch the horizon coincides with the size of the Metagalaxy in order of magnitude, then at the beginning of expansion, the former was much larger so that all regions of the Metagalaxy were causally interrelated.

In other words, the current value of the horizon corresponds to very close regions in the distant past. This is an exclusive feature of de Sitter's cosmology. It should be stressed once more that describing the initial stage of the expansion of the Metagalaxy with the help of de Sitter's model can also resolve the energy problem: the origin of the Metagalaxy (or of the world surrounding us) out of nothing is nonsense; this was always the opinion of the great philosophers and physicists. B. Spinoza was excommunicated by the orthodox Church in 1651 only because he doubted the biblical dogma of genesis out of nothing. Einstein always doubted the physical existence of a singular state, considering it just a consequence of mathematical idealization.

Introducing a physical vacuum as well as two stages in the evolution of the Metagalaxy can solve the basic problems of cosmology. Let us consider Fig. 4.3, illustrating a combination of a phase jump in a physical vacuum with a transition from the de Sitter stage to the Friedmann stage. One should note, however, that the outlined scheme of the phase transitions, albeit a most simple one, is not unique. Probably, in the first moments, the Metagalaxy went through a number of jumps before it acquired its features.

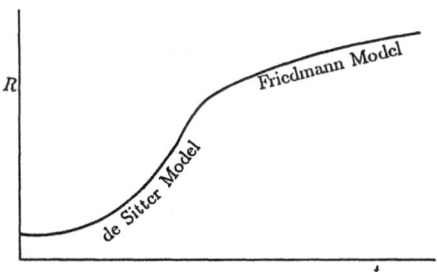

Fig. 4.3. Transition from the de Sitter to the Friedmann stage of the expansion of the Metagalaxy (schematic)

The present state of the theory of a physical vacuum and of the theory of gravitation does not give an unequivocal answer to the question of what was happening in the very first moments of the Metagalaxy. However, just the possibility of outlining the probable course of this evolution (or revolution) is inspiring.

We now pass to our main subject. It has been mentioned above that the baryonic asymmetry of the Metagalaxy arises at times of the order of 10^{-35} s. Phase "convulsions" in the Metagalaxy take place at still shorter times. The attempts to provide a physical description of the evolution of the Metagalaxy at these short times inevitably lead to the question: What was happening with the Metagalaxy at the moment of its birth, i.e., at $t \approx 0$? An answer to this question suggests the next one: What can be said about other regions of the Universe before the rise of the Metagalaxy, i.e., at $t<0$?

The birth of such a unique object as the Metagalaxy must be treated with the greatest respect; a description of this process requires a very cautious approach guided, as far as possible, by well-established physical facts. In the next two sections, such facts will be presented. We may appear to digress from our topic, but, as is well known, when dealing with difficult problems, a straight path is not always the shortest one.

4.9 The Structure of the Metagalaxy and the Fundamental Constants

A standard method of analyzing a physical phenomenon usually consists in repeatedly reproducing it. Unfortunately, the event of the birth of Metagalaxy cannot be reproduced under laboratory conditions. Furthermore, we cannot model extra-metagalactic objects (EMGO) experimentally, but there is no obstacle to doing so at a desk. In this matter, the standard methods appear to lead to a dead end; therefore, it is necessary to depart from familiar ways.

At present, a theoretical description of the birth of the Metagalaxy is not feasible. (The hopes of developing such a theory will be discussed later on). Hence we shall choose a different path: we shall portray the Metagalaxy under somewhat varied physical regularities. The result will be amazing: upon small changes, the Metagalaxy changes its appearance completely. This happens in one direction only, namely, always toward a simpler structure.

These conclusions involve certain notions requiring clarification. First of all, what is meant by a "small" variation of physical regularities and, second, in what sense is a "simpler" structure of the Metagalaxy to be understood?

Physics is a combination of the physical laws (such a Newton's law, Coulomb's law, etc.), the concepts of physical space (Euclidean, Minkowski, Riemann), and the values of the fundamental constants whose set is non-conventional. We shall consider as the fundamental constants the interaction constants of four interactions as well as the proton mass, m_p, the neutron mass, m_n, the electron mass, m_e, and the dimension N of physical (conceptual) space. From among the multitude of elementary particles, three (viz., the proton, the neutron, and the electron) have been chosen because the Metagalaxy is composed predominantly of these three kinds of particles in combination.

Now it is necessary to clarify the meaning of a "small" variation. A change in the laws of physics (such as, e.g., replacing the Coulomb law $F \sim r^{-2}$ by, say, $F \sim r^{-3}$) would lead to a dramatic change. Altering the structure of physical space would likewise have revolutionary consequences. Hence, when referring to small variations of physical regularities, we shall mean a (relatively) small change in the fundamental constants, the physical laws and the structure of physical space being invariable.

Let us explain the notion of simplifying the structure of the Metagalaxy. The four major structural features of the Metagalaxy are the atomic nuclei, the atoms, the stars, and the galaxies. (The so-called large-scale structure of the Metagalaxy will not be considered here since in the modern picture, it is not regarded as a separate structural constituent of the observable world but is, rather, inherent in it as a whole). Simplification of structure is to be understood as the (imaginary) disappearance of one or several structural elements. (The arguments for the instability of the structure of the Metagalaxy with respect to the magnitude of the fundamental constants are given in Chap. 3.)

4.10 The Metagalaxy as a Fluctuation

When speaking of fluctuations, one commonly means the distribution of a certain quantity whose particular magnitude strongly deviates from its average value. There is no set of EMGO's available to study their distribution. However, we have access to extremely well-studied distributions of the quantities which, as seen in the preceding section, determine the structure of the Metagalaxy.

An unexpected and extremely important conclusion then follows: The fundamental constants of an object that give rise to the occurrence of a complex structure are fluctuations in the distribution of the corresponding parameters (e.g., of the mass) of similar objects. A very simple example is the mass spectrum of the elementary particles. Fig. 4.4 shows the mass distribution of the elementary particles. To date, several hundred elementary particles have been investigated experimentally, their masses being distributed over four orders of magnitude (ranging from the electron mass of 10^{-27} g to the Ψ-particle mass of 10^{-23} g). A convenient representation of this distribution is therefore in the semilogarithmic scale. A sharp maximum near the proton mass m_p can be recognized. About 90 % of all particles have a mass coinciding with the proton mass to within a factor of 2. Only several particles have a mass differing from m_p by a factor of about 10 (the muons and the Y-particles).

From an analysis of Fig. 4.4, one can readily conclude that compared to other masses, the electron mass is so extremely small that a minute increase of it would result in neutronization.

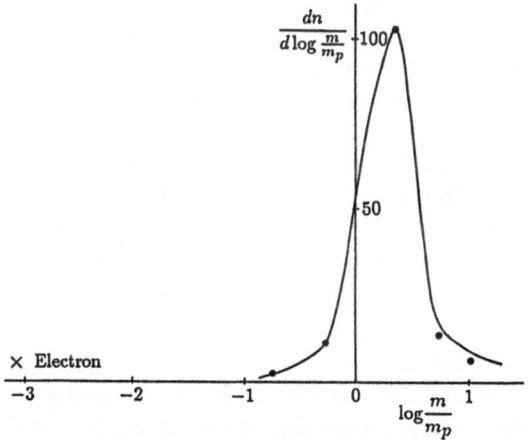

Fig. 4.4. Mass distribution of the elementary particles. On the abscissa, the logarithm of the elementary particle mass is given in units of the proton mass. The ordinate represents the ratio of the particle number dn to the "logarithmic mass" $d\lg(m/m_p)$ (mass in the units of the proton mass), i.e., the particle number per unit "logarithmic mass"

One can attempt to quantitatively estimate the probability of a fluctuation in the actual mass distribution of the particles, i.e., the probability of the occurrence of a particle with mass within the interval $(1-3)m_e$. To that end, the experimental mass distribution will be approximated by simplest power laws. The broken line in Fig. 4.4 is the result of such an approximation: for the elementary particle mass $m > 2m_p$, one has for the distribution $\Delta n / \Delta \lg (m/m_p) \approx 300(m/m_p)^{-1.5}$, while for $m < 2m_p$, the distribution obeys the relation $\Delta n / \Delta \lg (m/m_p) \approx 30(m/m_p)^2$. Accordingly, using the approximation proposed, one can evaluate the probability of the occurrence of a particle with a mass close to m_e. This probability turns out to be below 10^{-5}.

We have proved earlier that a small increase of the magnitude of m_e would result in a major change of the structure of the Metagalaxy. But the very value of m_e observed is a large fluctuation in the mass distribution of the elementary particles. If one assumes the existence of metagalaxies similar to the Metagalaxy in all but the value of m_e (and having, instead, a mass equal to the average particle mass), then the Metagalaxy can be said to be one fluctuation among others.

Let us take another example. In elementary particle physics, the notion of isospin multiplets is of great significance (cf. Sect. 3.2.3). Isospin multiplets are characterized by identical quantum numbers (except for the value of the electric charge) and by almost the same mass. The mass difference within a multiplet is not more than 0.1–1 %. A typical example of isospin multiplets are the nucleons: the protons and the neutrons.

The electric charge of a proton is equal to that of an electron, while the electric charge of a neutron is zero. The mass difference between the two particles is about 0.1 %, all other quantum numbers being identical. It is the small mass difference which gives rise to the stability of a deuteron and to the existence of complex chemical compounds (cf. the preceding section). Such a small mass difference is not representative of other isospin multiplets, however. Figure 3.2 shows the distribution of the isospin multiplets over the mass differences. It is recognized that the mass difference $\Delta m_N = m_n - m_p$ is significantly smaller than a similar quantity for other isospin multiplets. Moreover, were the mass difference Δm_N a minimum among all such mass differences determined (leaving aside $m_{\Sigma^0} - m_{\Sigma^\pm}$, the mass difference between the Σ^0-particle and the $\Sigma^\pm$-particle), the deuteron would be an unstable particle.

Finally, a last example. As mentioned in the foregoing section, a very small increase of the strong interaction constant α_s (or of the deuteron binding energy) would result in the occurrence of a stable bi-proton. The deuteron binding energy ε_d is extrememly small as compared to the corresponding quantity for other nuclei, however. In nuclear physics, it is customary to base a comparison on the binding energy per particle (nucleon) in the nucleus. It turns out that for a deuteron, this quantity is of the order of 1 MeV, whereas for other nuclei it is about 8–9 MeV, an exception being the Li nucleus, where this quantity has the value of about 6 MeV.

4.11 The Anthropic Principle

The Church contended against the teaching of Nicolaus Copernicus because it contradicted theological doctrine. The logic of the clergy was that if the leader of the Israelites, Joshua, the son of Nun, had ordered the Sun to stand still[10], then it must have moved prior to his command. But Copernicus asserted in contradiction to the Bible that the Sun is immovable. The real issue was actually the more essential thesis declared by the Church that man is the lord of creation; to replace the statement "Earth is the center of the Universe" by "Earth is an ordinary planet" was to cast doubt upon this paramount biblical thesis. It was not possible to reconcile these two conflicting statements and thus the teaching of Copernicus was anathematized.

However, as often happens in the heat of an argument, in the process of establishing the new "Weltanschauung", certain extreme tendencies – of which one was not aware until recently – became prevalent. They found their expression in the so-called anthropic principle (cf. Sect. 3.6). To understand its essence, we shall try to concisely analyze the ideas of Copernicus.

4.11.1 Definition

Of course, Earth is an "ordinary" planet, nonpreferential with respect to the laws of mechanics. Although "ordinary" with regard to physics, this celestial body is unique in one respect: it is the only known dwelling-place of Man, whose intellect is capable of apprehending the Metagalaxy. One can endlessly dispute the existence of extraterrestrial civilizations, but it is certain that Man appeared on Earth several hundred thousand years ago, and that the Mesopotamian civilization is about 10 thousand years old – an age infinitely small compared with the age of the Metagalaxy. Investigations with cosmic apparatuses have also demonstrated unequivocally that there are no signs of organic life on the Moon, Venus or Mars.

Hence, we have to recognize that the occurrence of a being able to perceive Nature imposes certain conditions on space and time. The development of the human mind was preceded by a long evolution of physical factors in

---

[10] Josh. 10:12, 13

the Metagalaxy; it appears very probable that its existence would be impossible without the appearance of galaxies and stars which were formed after the Metagalaxy had been inflating for more than one billion years.

One may state with some certitude that in a way, the human mind is the crown and lord of "Nature's creation." This is the idea underlying the anthropic principle. It has been formulated in a most pronounced form by the British astrophysicist B. Carter. Paraphrasing the famous saying by Descartes "Cogito, ergo sum" ("I think, therefore I am"), he suggested the formula "Cogito, ergo mundus talis est" ("I think, therefore the world is what it is").

Of course, it is not good practice to base physical conclusions (to be discussed below) on such issues as the beginning of life, which are rather ill-defined and distant from physics. The author would prefer a less elegant but more strict definition of the principle. There is, for example, a principle according to which optimal conditions are required for complex forms of matter to occur. One of these conditions is, primarily, the sufficiently long existence and evolution of stars, galaxies, and the Metagalaxy. The statement "The physical laws governing the Metagalaxy are sufficient for the rise of life" is trivial.

However, another statement is nontrivial: "These laws are necessary for the rise of complex forms of matter." One should note the profound expediency and harmony of the physical laws; this also applies for the values of the fundamental constants: a minute variation of physical regularities would render the existence of complex forms of matter impossible.

4.11.2 Applications

Let us turn our attention to some applications of the anthropic principle, and consider the interpretation of certain cosmological coincidences of numbers that have been intriguing many investigators for decades. For instance, it has been known for a long time that the average density of matter ϱ in the Galaxy is close to the critical density $\varrho_c \approx 10^{-29} \, \text{g/cm}^2$. (It should be recalled that the Metagalaxy expands unlimitedly if $\varrho > \varrho_c$ and contracts otherwise). Why, of all the possible densities, this is one realized in the Metagalaxy?

An answer based on the anthropic principle is that were the inequality $\varrho \ll \varrho_c$ fulfilled in the Metagalaxy, it would follow from calculations that the relative velocity of its particles would be too high for diffuse matter to condense into galaxies and stars. In the opposite case of $\varrho \gg \varrho_c$, the condensation process would be incomparably easier than in the real Metagalaxy. Its lifetime, t_M, would then be short, however ($t_M \sim \varrho^{-1/2}$). Due to this shortness, life would not have a chance to develop in such a metagalaxy.

Hence, for the Metagalaxy in which complex forms of matter do exist, the condition $\varrho \approx \varrho_c$ must hold.

Let us consider another related example. As has already been mentioned, the density of a physical vacuum is small ($<10^{-29} \, \text{g/cm}^3$). This figure is not just small: it is inconceivably small compared to the terrestrial scale and also to estimates based on dimensionality considerations. With the aid of the anthropic principle, the problem can be solved by reasoning similar to that applied in the above example. Since a physical vacuum is one of the forms of matter, it must gravitate, like any material medium. If the inequality $\varrho > 10^{-29} \, \text{g/cm}^3$ held, then the lifetime of the Metagalaxy would be too short to permit the development of beings with intellect.

Still another example of the application of the anthropic principle can be described. For a long time (beginning with Dirac's well-known paper published in 1937) physicists have been concerned with the extreme smallness of the nondimensional gravitation constant α_g which is of the order of 10^{-39}. This quantity is incredibly small compared with the electromagnetic interaction constant $\alpha_e \approx 1/137$. Why is the nondimensional gravitation constant so small?

Up to now, the only satisfactory answer to this question has been provided by the anthropic principle. As mentioned above, the lifetime of the stars of the main sequence, including our Sun, is proportional to α_g^{-1} (cf. (3.16)). Hence, if, for instance, the magnitude of α_g increased by two orders of magnitude, the lifetime of the stars and of the Sun would decrease by two orders of magnitude, and life would have no chance of arising during this time.[11]

Thus, some successful applications of the anthropic principle have been demonstrated. In the author's option, speculation about the relationship between the "parameters" of human beings and the parameters of atoms and planets is very doubtful. For example, it is posulated by some authors that the mass and the dimensions of Man are equal to the geometric average of the corresponding parameters of Earth and the atoms. The geometric average of the mass is then about 100 g and of the size, about 10 cm. Taking the geometric average of the parameters of the Sun and the atoms leads to about 100 kg and 100 cm, respectively. This is quite close to the realistic values.

Why is it the Sun which has to be taken as a macroscopic entity, though? Indeed, it is evident that the gravitation of the Earth had the larger effect on biological evolution. Of course, the idea of the "parameters" of living creatures being some average of the micro- and macroparameters is quite appealing. Nevertheless, a simplistic quantitative realization of this idea by evaluating the geometric average appears to be too superficial.

---

[11] Another example for the application of the anthropic principle can be found in Sect. 3.6.

4.12 The Birth of the Metagalaxy and of Metagalaxies

Let us summarize what has been said about the Metagalaxy:

1. The complex structure of the Metagalaxy is very unstable with respect to the numerical values of the fundamental constants. A slight variation of these quantities would result in a significantly simpler structure.

2. The magnitude of these constants as found in the Metagalaxy suggests that they are large fluctuations in the distributions of the experimentally studied constants which characterize related objects (metagalaxies).

3. The magnitude of the fundamental constants remained unchanged during most of the evolution of the Metagalaxy.

4. The Friedmann model that provides a description of almost the whole cycle of the expansion of the Metagalaxy encounters a number of difficulties when dealing with the initial stage, i.e., with time $t<10^{-35}$ s.

5. At $t<10^{-35}$ s, a phase transition from de Sitter's cosmology to Friedmann's cosmology probably occurs.

6. In the process of this phase transition, a re-structuring of the physical vacuum takes place. As a result, the (presumably) large energy density ε_{vac} takes a small, or maybe even zero, value.

7. Many an enigmatic cosmological fact is accounted for by the assumption that the Metagalaxy followed a course of evolution having very definite values of the fundamental constants (the anthropic principle).

How can all these facts be arranged into a coherent picture?

In principle, one may assume, as did Dirac, that the fundamental constants vary during the evolution of the Metagalaxy. We are living in an epoch of a favorable combination of the fundamental constants. However, the experimental data (item 3) contradict the Dirac hypothesis. Besides, this hypothesis does not resolve the difficulties of the Friedmann model. Therefore Dirac's hypothesis should be abandoned.

One could assume (as was done inadvertently for a long time) that the Metagalaxy, with all its physical laws and fundamental constants, formed at the moment of singularity ($t = 0$). As far as pure logic is concerned, such a picture does not directly contradict the above items, except for the assumption of the existence of an appropriate de Sitter cosmology. However, in adopting this picture we have to imply a practically improbable coincidence of lucky circumstances. In the first place, this refers to the favorable combination of the fundamental constants whose estimated probability is extremely small. Secondly, all the difficulties of Friedmann's model remain unresolved. Thirdly, the very picture in which there was nothing at $t<0$, while after $t = 0$, an object like the Metagalaxy exists, is in conflict with our entire experience and, possibly, with the law of conservation of energy.

It appears that, instead of adopting a cosmology which assumes the existence of a single Metagalaxy with invariable physical laws, it is less revolutionary and also most simple to assume that the Metagalaxy is a closed system (implying that the average density of matter is supercritical, $\varrho > \varrho_c$) and that it undergoes many evolutionary cycles (Fig. 4.5). The instant $t = 0$ is then just the beginning of a cycle rather than the moment of the rise of the Metagalaxy.

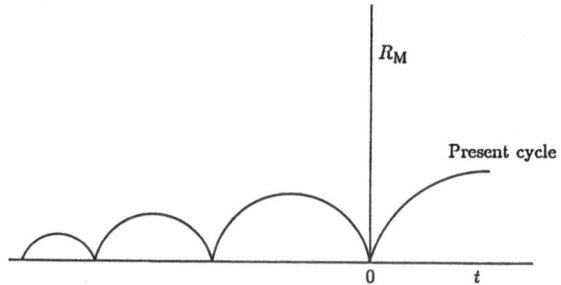

Fig. 4.5. Dependence of the quantity R_M on the cosmological time t for an oscillating Metagalaxy

Each advance to the next cycle involves a change in the set of the fundamental constants, including the energy density of the physical vacuum, ε_{vac}. We happen to live in a cycle in which conditions occur favoring the generation and the existence of organic matter. At $t < 0$ the Metagalaxy went through another cycle, with different values of the fundamental constants, and had a different, much simpler structure.

Such a picture, while accounting for almost all facts listed above, still has certain deficiences. First, a model of an oscillating Metagalaxy implies that its mass increases with each cycle. Consequently, there must have been a moment when the Metagalaxy had an infinitesimal mass. In other words, introducing the notion of an oscillating Metagalaxy again leads to the conclusion that it emerged out of nothing.

Second, when treating the fundamental constants and their possible changes, we did not discuss the dimension of physical space. It has been mentioned earlier, however, that the value $N = 3$ is a very special one. It is the maximum value of the dimension N which admits of the existence of atoms, planets, or analogues thereof. The oscillating Metagalaxy model, for example, implies that the dimension N has always been invariable, $N = 3$. It turns out, then, that again we are very lucky: of all possible values of the dimension of space ranging from unity to infinity, Nature chose the only one which permits complex forms of matter to occur.

There are also some further problems with the oscillatory model which have not been overcome as yet. Thus, the picture of the oscillating Metagalaxy resolves many problems and is appealingly simple; it does meet with some difficulties, however, and this has to be taken into account when analyzing principal questions of cosmology. It is useful therefore to search for other, more complex approaches to solving them.

4.12.1 Progress in Cosmology Brings Understanding

To comprehend the ideas presented below, it is worth-wile considering a primitive analogy which is quite illustrative. Supposing we are interested in the fate of a stone thrown up with an arbitrary velocity. Newtonian mechanics makes it possible to calculate the trajectory of the stone with all the precision required. This theory is absolutely useless, however, if the fate of the stone after its fall is to be predicted. This fate is determined by the properties of the material hit by the stone rather than by the mechanics of the fall. If the stone falls, e.g., onto sufficiently soft ground, it will remain at the point of impact. If it falls into a swamp or into water, it will be out of sight for the observer altogether. If it falls onto an elastic film, it will be reflected and will fly along a trajectory which depends on the properties of the film.

Until recently, cosmologists were in a situation similar to that of a person observing the motion of a stone without knowing the properties of the ground on which the stone falls. Cosmology was based, almost exclusively, on the theory of gravitation, i.e., the general theory of relativity. The entire evolution of the Metagalaxy as a whole was considered to be determined by gravitation. The origin of the Metagalaxy as well as its "destination" after the completion of a cycle did not interest cosmologists. This was quite natural, for the statements presented at the beginning of this section were unknown or went unheeded.

Let us take into account the progress made in the understanding of the nature of the physical vacuum and the fundamental constants. Let us further assume that the dimension $N = 3$ is one of the constants characteristic only of the current cycle of the evolution of the Metagalaxy, which is capable of giving rise to life. Then the following interrelated questions arise immediately: What came first, a physical vacuum or the Metagalaxy? Does anything exist beyond the Metagalaxy?

If one can transcend the mind-boggling scale of the Metagalaxy, it is quite reasonable to assume that beyond it, metagalaxies as well as a physical vacuum exist. This picture is not an idle fantasy of the author; many cosmologists model — explicitly or implicitly — the formation of the Metagalaxy from a physical vacuum. Owing to many uncertainties caused mainly by our limited knowledge of the physical vacuum, such models are of course not unequivocal. However, at the time this chapter is being written, the trend of linking the formation of the Metagalaxy to the properties of the physical vacuum is quite clear.

A somewhat novel problem here is the question we have touched upon repeatedly, viz., that of the dimension of space. But while this might be new to cosmologists, it is a familiar and fundamental concern of physicists specializing in the theory of unified interactions.[12]

In the last decade, the so-called theory of supergravity unifying all four basic interactions, including the gravitational interaction, has made substantial progress. This theory strives to eliminate singularities (divergences) arising when calculating the characteristics of the interactions. It has turned out, however, that constructing a consistent theory of supergravity within three-dimensional space is a very difficult or maybe even impossible task. The smallest dimension for which the divergences can be overcome is $N = 10$. (This does not take into account the time coordinate, so that $N = 11$ would correspond to the Minkowski space.)

This idea, which is geometric in its essence, can be made use of in cosmological considerations. Currently, Kasner's so-called cosmological solution is a subject of vigorous discussion. It has a very interesting property: for a space of dimension N, this solution decreases with time for a subspace of dimension N' and increases for the rest of $N-N'$ coordinates.

This peculiar cosmological solution to the general theory of relativity can be given a different interpretation. In cosmological approaches, deforming coordinate systems are often used, deformation being in one direction only. Applying the principle of deforming coordinate systems to Kasner's solution would, by analogy, mean that part of the coordinate system (dimension N') deforms in the opposite direction, namely, toward lower values of the coordinates. Thus, for a sufficient degree of deformation of the coordinate system, compactification would occur, i.e., a practical reduction of the system of N coordinates to that of $N-N'$ coordinates.

Based on this principle, a model of a multidimensional space (e.g., with $N = 10$) can be developed which, in the course of time, reduces to three-dimensional space. The rest of the space dimensions can only come to bear at a scale below 10^{-32} m. Hence, the use of such a multidimensional space does not contradict the fact that no manifestation of the compactified dimensions of the space is detectable.

In a theory of this kind, the formation of objects like the Metagalaxy can be conjectured as follows. A multidimensional background space exists, filled with a physical vacuum subject to perturbations. These perturbations

---

[12] T. Kaluza was the first to raise, in 1921, the question of the relation between the unification of interactions and the dimension of the space involved.

give rise to the evolution of objects like the Metagalaxy. It is conceivable, for example, that such a perturbation stems from a merger of some metagalaxy with the general background (at the end of cosmological contraction, say).

In the process of the development of this perturbation, compactification of the dimensions associated with the background space may have occurred, either accidentally or due to a certain regularity. Then the disturbance would have developed, first according to the de Sitter model and subsequently, following the Friedmann model. In particular, in the course of this development the set of the fundamental constants that determines the complex structure of the Metagalaxy would have been realized[13].

In the light of the above, the following picture of the entire Universe and its evolution can be sketched. The Universe – eternal and infinite – lives a stormy life reminiscent (metaphorically, of course) of a pot of boiling liquid. Like vapor bubbles, metagalaxies arise, expand, and finally die, giving birth to new metagalaxies. Such a picture seems to be very natural; all the problems and difficulties outlined in the beginning of this section can be resolved on this basis.

Last but not least, one may ask whether it is possible in principle to check on the validity of this scenario of the birth of the Metagalaxy, in the near future, if not now. Is it not pure fantasy?

As far as a direct experimental verification is concerned, no approach is in sight. However, currently there are serious hopes that the advancing development of the theory of supergravity and of the theory of the physical vacuum as well as their synthesis will make it possible to solve the problem of the birth of the Metagalaxy. And it is but the outline of this grandiose phenomenon that we now see, though not very distinctly yet, much as we might see the silhouette of a large building against distant lightning in the night.

4.13 Future of the Metagalaxy

Modern astrophysics and cosmology have reached a level that justifies the attempt to predict the character of the future evolution of the Metagalaxy. This is true, but with some reservations. First, this evolution is uniquely determined by ϱ, the average density of matter in the Metagalaxy. If $\varrho < \varrho_c$ (i.e., if the Metagalaxy is open), some aspects of its evolution differ markedly from the development of a closed Metagalaxy. Second, in describing the end of a cycle of an oscillating (closed) Metagalaxy, problems similar to those encountered earlier, when analyzing the process of the birth of the

---

[13] This concept, somewhat modified, has been pursued in the works by A.D. Linde, S. Hawking, the author, and other physicists.

Metagalaxy (at $t = 0$), arise. Unfortunately, the available experimental data do not provide an answer to the question of the relative magnitude of ϱ and ϱ_c. Direct observations have demonstrated that the average density of luminous matter, ϱ_{lum}, amounts to about $0.1\,\varrho_c$.

One might readily conclude from this fact that we are dealing with a model of an open Metagalaxy. However, this conclusion would be premature: other observations indicate that the Metagalaxy contains invisible, "hidden" matter whose density may exceed ϱ_{lum} by a factor of 10 to 100. It is possible, therefore, that $\varrho > \varrho_c$. Consequently, two options have to be considered, namely, an open and a closed Metagalaxy (cf. Sect. 2.4).

4.13.1 Open Metagalaxy

In Friedmann's cosmology, an open Metagalaxy has to evolve eternally. As evolution goes on, the volume of the Metagalaxy will tend to infinity and, accordingly, the average density of matter and radiation will go to zero. However, the concept of the Metagalaxy tending to infinite inflation must be used with caution. Traditional cosmology recognizes unlimited expansion as a perfectly natural process; however, as seen above, it is quite probable that other objects besides the Metagalaxy exist. Then interactions between the Metagalaxy and metagalaxies are practically unavoidable.

The simplest physical analogue of such "collisions" are the collisions of vapor bubbles in a fluid: Two bubbles that expand and move in the fluid may collide. If the Metagalaxy expands infinitely, and if other objects are also present, then such interactions are practically – but not theoretically – inevitable. One can imagine, for instance, that metagalaxies expand in different dimensions (i.e., in different subspaces), thus making collisions impossible. This possibility, although conceivable, seems to be rather improbable.

At present, there is no model available to describe these interactions. It is natural, therefore, to remain within the framework of the canonical model if one wishes to forecast the evolution of metagalaxies, i.e., to consider the Metagalaxy to be open and expanding infinitely. Let us consider, in this case, the fate of the major elements of the Metagalaxy (i.e., of the stars and galaxies). We shall assume, of course, that the physical laws remain unchanged throughout the further evolution of the Metagalaxy.

All elements of the Metagalaxy are destined to gradually cease their activity. This process consists in the cooling down of the objects which sometimes can be very slow, but sometimes very rapid. The cause of this phenomenon is simply the law of conservation of energy. Cosmic objects are radiating, while the energy stored in them is finite. Consequently, sooner or later the stored energy will be exhausted and the objects will turn into huge frozen blocks, wandering in the boundless Metagalaxy.

Let us consider some particular processes in the evolution of celestial bodies, primarily stars. Let us recall that at present, four basic classes of quasi-stationary stars[14] have been observed: the stars of the main sequence, the white dwarfs, the neutron stars, and the red dwarfs (cf. Sect. 2.9). The Sun is a typical star of the main sequence, and its activity (radiation) is maintained at the cost of thermonuclear energy released as a result of collisions between the nuclei of light elements contained in its depth.

It should be noted that the first three of the above classes of stars are genetically connected. When thermonuclear energy stored in the stars of the main sequence gets exhausted, their precipitous contraction (collapse) gives rise to white dwarfs, neutron stars, or black holes (which will be considered below).

The lifetime of the stars of the main sequence is also determined by their mass. The stars with a relatively small mass (e.g., the Sun) have an intense brightness for a long time. The Sun, for example, will not significantly change its characteristics for at least 10 billion years; in 10–20 billion years it is destined to transform into a white dwarf. Stars more massive than the Sun radiate much more and, consequently, stay in the main sequence for only a relatively short time. For example, stars with a mass of about $10M_S$ will leave the main sequence in approximately 100 million years.

The further fate of white dwarfs and neutron stars is determined by their luminosity, as well as by the kinetic and the thermonuclear energy of the particles of which these objects are made up. The lifetime of white dwarfs in the luminous state is of the order of 10^{14} years. Afterwards, they will turn into dark blocks whose temperature is about 1 K. Neutron stars have the same future, only their cooling time is longer, 10^{19} years, and their final temperature is about 100 K.

A somewhat peculiar fate awaits stars with a small mass, in particular, the red dwarfs. Red dwarfs possess two noteworthy features: they radiate much less than the stars of the main sequence (by a factor of 10^3–10^4 less than the Sun) but their mass is also substantially smaller (about $0.05\,M_S$). Consequently, the radiation of the red dwarfs is caused not by thermonuclear reactions (as in stars like the Sun), but rather by gravitational energy. It turns out that this energy will suffice for about 100 billion years. Then, after a comparatively sluggish collapse, the red dwarfs will transform into something like small white dwarfs.

The existence and evolution of stars with a very small mass ($0.001\,M_S < M < 0.004\,M_S$) is very intriguing. Our knowledge of them is, however, wanting, since it cannot transcend the observability limit, represented by the mass of the red dwarfs: $0.04\,M_S$. But stars with still smaller masses radi-

---

[14] Objects whose lifetime is comparable with the time of the existence of the Metagalaxy are referred to as quasi-stationary stars.

ate so weakly that they cannot be resolved by the devices available. For example, the luminosity of the largest planet, Jupiter $m_{\text{Jup}} \approx 0.001\, M_S$, is about 1 billion times smaller than the luminosity of the Sun. It is not feasible to observe such "star planets" at distances exceeding one parsec. Stars with very small mass radiate very weakly, become extinct very slowly, and, finally, develop into cold, solid bodies without experiencing any cataclysms.

The galaxies have a very curious fate. Their quasi-stationary nature results from a balance between the attractive gravitational forces and repulsive centrifugal forces. In some rare cases, this balance can be broken, however. This happens if one of the stars has a velocity substantially exceeding the average velocity of the stars in a given galaxy. The simplest analogue of such a process is the slow evaporation of water at room temperature: the water molecules with the largest velocities gradually leave the vessel, slowly draining it.

On a cosmic scale, stars are similar to molecules, and they leave their "vessel," the galaxy. This "evaporation" process can be evaluated on the basis of the observational data on the characteristics of the galaxies and their constituent stars. It proves that the "evaporation" of stars from the galaxies will cause erosion within a time of the order of 10^{18}–10^{19} years. Collapse of the galaxies will be a consequence.

Up to now we have based our considerations on physical laws, reliably established in laboratory tests, using their extrapolations in space and time. Prediction of the further fate of the Metagalaxy (at times exceeding 10^{20} years) depends on theoretical speculations which have not been confirmed by experiment as yet. We shall refer to concepts currently shared by the majority of specialists. Unfortunately, the fundamental conclusions we shall draw cannot be validated by reliable experimental data.

In the first place, proton decay should be considered (cf. Sect. 1.6). It is well known from experiment that a proton is a stable particle in the sense that its lifetime is longer than the time the Metagalaxy has existed. The modern theory of unified interactions predicts, however, that a proton must decay, its lifetime being $10^{31\pm2}$ years. Should this prediction be correct, almost all of matter will consist, after a time of the order of 10^{31} years, of electrons, positrons, photons, and neutrinos, scattered in space with extremely low density.

Note the above proviso: "almost" all of matter. Will anything remain in such a decaying world? The anwer is: Yes, black holes. These objects, also created by the fantasy of the theoreticians, are capable of absorbing nearly all particles and quanta encountered in their path. Except for the quarks, black holes are unrivaled in their popularity as theoretical objects yet to be reliably observed. The properties of black holes have been repeatedly discussed, and it does not seem to be worth-while dwelling on their characteristics.

One feature of the black holes should be mentioned, however; it was discovered (at the tip of a pen, of course) some ten years ago. Black holes are not absolutely frozen objects as was supposed before; they also must continually use up their mass to radiate photons and particles. However, the lifetime of a black hole with a mass of the order of the mass of the Sun is extremely large, 10^{70} years. After the evaporation of the black holes due to radiation, the Metagalaxy will consist exclusively of light particles.

4.13.2 Closed Metagalaxy

The future of a closed Metagalaxy appears to be somewhat different. To a great extent, the difference stems from the fact that the characteristic time of the current cycle of the Metagalaxy is comparable with or smaller than the characteristic times of the above-mentioned processes. Furthermore, a closed Metagalaxy does not inflate infinitely, on the contrary, upon reaching a certain maximum size, $R_M \approx R_{max}$, it will start contracting down to a very small size. It should be noted once again that, in accordance with the anthropic principle, we probably live in an epoch close to a "golden age" of the Metagalaxy, when $R_M \approx R_{max}$.

The age of the closed Metagalaxy, short compared with the times characteristic of the extinction of many cosmic objects (dwarfs and neutron stars), will not allow such objects to simply "die peacefully." Turbulent processes accompanying the swift contraction of the Metagalaxy in its final stages will raze them to the ground.

Let us reckon the time t_1 backward from the end of the collapse of a closed Metagalaxy. Then, at $t_1 \approx 10^6$ years, the remaining stars of the main sequence will start decaying. At $t_1 \approx 100\,\mathrm{s}$, the white dwarfs and at $t_1 \approx 10^{-4}\,\mathrm{s}$, the neutron stars will be destroyed, respectively. At that time, the Metagalaxy will consist of loose protons, neutrons, electrons, photons, neutrinos, and black holes.

Whereas up to this point one could sketch the scenario of evolution with the firm strokes of a paintbrush, the further fate of the closed Metagalaxy (especially at $t_1 \to 0$) appears more nebulous. It is likely that at $t_1 \lesssim 10^{-35}\,\mathrm{s}$, a certain number of free particles will exist, including superheavy ones, in addition to the black holes. This entire conglomeration will fuse to form something like a huge black hole. Traditional cosmology further envisions a subsequent collapse leading to a singular state.

In our opinion, it is more consistent to take a viewpoint related to the concept of the beginning of the Metagalaxy. That is to say, at very small times, $t_1 \lesssim 10^{-35}\,\mathrm{s}$, phase transitions leading to a merger of the Metagalaxy with the physical vacuum ought to occur. This process may give rise to the formation of new metagalaxies. It is reasonable to assume that these phase transitions are accompanied by an exchange of the (Friedmann-type) power-law contraction of the Metagalaxy for the exponential one, which corresponds to de Sitter's cosmology. The end of the Metagalaxy is, in a way, a mirror reflection of its beginning.

It should be emphasized once again that we are dealing here with a scenario of a sequence of physical pictures. "Scenario" is a favorite word used by cosmologists when analyzing issues connected with the beginning and end of the Metagalaxy. Even though a scenario is far from being a produced movie, it appears that we have made significant advances in our understanding of the qualitative picture of the beginning and the end of the Metagalaxy. Our knowledge is not sufficient to quantitatively evaluate the details of this picture, however.

This quantitative evaluation is the primary goal of the quantum theory of gravitation, which is expected to be the most important element of a theory that will unify all interactions. Most momentuous and promising in this respect is the rapprochement of the fundamental disciplines cosmology and elementary particle physics, as well as progress in the understanding of the nature of the space dimensionality and the fundamental constants.